DISTANCE IN GRAPHS

DISTANCE IN GRAPHS

FRED BUCKLEY

Baruch College,
City University of New York

FRANK HARARY

New Mexico State University

ADDISON-WESLEY PUBLISHING COMPANY

The Advanced Book Program

Redwood City, California • Menlo Park, California • Reading, Massachusetts
New York • Don Mills, Ontario • Wokingham, United Kingdom • Amsterdam
Bonn • Sydney • Singapore • Tokyo • Madrid • San Juan

Publisher: *Allan M. Wylde*
Production Manager: *Jan V. Benes*
Electronic Publishing Consultant: *Mona Zeftel*
Promotions Manager: *Laura Likely*
Cover Design: *Iva Frank*

Library of Congress Cataloging-in-Publication Data

Buckley, Fred.
   Distance in graphs / Fred Buckley, Frank Harary.
     p.   cm.
   Bibliography: p.
   Includes index.
   1. Graph theory.   I. Harary, Frank.   II. Title.
QA166.B83   1989   511′.5—dc19   89-245
   ISBN 0-201-09591-2

This book was prepared by the authors, using T$_E$X typesetting language on an IBM computer.

BCDEFGHIJ-MA-9543210

Dedicated to

FRANK BOESCH
and
PETER SLATER

Two of the most outstanding pioneers
on distance concepts in graph theory

Preface

Graph Theory has developed into a very active area of mathematical research. Whereas twenty years ago many mathematics departments had no graph theorists, it is now not uncommon to find several in a single department. A major impetus for this growth has certainly been the wide applicability of graph theory, especially in computer science. Besides the dozen or so extant introductory texts on Graph Theory, monographs have been written in recent years covering such specialized areas as connectivity, colorability, extremal graphs, random graphs, ramsey theory, and groups and surfaces. A number of recent introductory texts have used algorithms as the common thread, a reasonable approach because of the interest of computer scientists in graph theory. One concept that pervades all of graph theory is that of distance. Distance is used in isomorphism testing, graph operations, hamiltonicity problems, extremal problems on connectivity and diameter, and convexity in graphs. Distance is the basis of many concepts of symmetry in graphs. The important application of facility location on networks is based on various types of graphical centrality, all of which are defined using distance. Many graph algorithms depend on the idea of finding collections of long paths within a graph or network. Since distance is such a pervasive notion in graph theory, the time has come for a text focusing on distance in graphs.

Distance in Graphs is based on the classic *Graph Theory* by F.H. and brings *Graph Theory* up to date on the topics covered. This text can be used by advanced undergraduates and beginning graduate students in mathematics and computer science and also as a comprehensive reference work for researchers in graph theory, communication networks, and the many fields using graph theory in applications. The basic mathematical background required here is "mathematical sophistication," although a good one-year course in discrete mathematics (or its equivalent) would be helpful. We have made every effort to see that concepts are carefully explained and motivated. Accompanying figures are used to elucidate and illustrate concepts throughout the text. Clear, illuminating proofs for all major theorems are given. Exercises are included to further clarify the material from each section. Many of the exercises contain results extending the concepts discussed in the section and cite the original source to provide a direction for further reading.

It should be noted that in the Exercise Sections, a simple statement without the words "show that" means to prove it anyway.

The text material is divided into two parts. We discuss Graphs and Digraphs in Part I, which contains ten chapters. Chapter 1 presents the basic concepts of graph theory. Chapter 2 focuses on Centers and related distance concepts used throughout the book. It is expected that the chapters will be read in the order in which they appear. However, only Chapters 1 and 2 are essential prerequisites to a full appreciation of the results in subsequent chapters. In Chapters 3 and 4, we study Connectivity and Hamiltonicity including the recent distance-related advances in these areas. Next we examine extremal distance problems such as radius and diameter minimal and critical graphs, detours, and long paths in graphs. In Chapter 6 we discuss matrices including many results on distance matrix problems. Chapter 7 considers various convexity concepts for graphs including graph metrics, geodetic graphs, and distance-hereditary graphs. Symmetry is the subject of Chapter 8 with particular attention to distance-related results. Over the years many graphical sequences, almost all of which are distance-based, have been introduced to study various graph properties. We give a thorough treatment of Distance Sequences in Chapter 9. Digraphs are studied in Chapter 10, where we describe extensions of the results in earlier chapters to digraphs.

Graph algorithms concerning distance are separated into Part II of *Distance in Graphs* so they may easily be incorporated into or deleted from a course depending on the background of the students. All algorithms are described in Pascal-like pseudocode. Chapter 11 gives numerous distance-related graph algorithms, while Chapter 12 focuses on distance algorithms in Networks.

The numerous references to appropriate journal articles given throughout *Distance in Graphs*, make it ideally suited both as an advanced book in mathematics and an important reference work in graph theory. Following each item in the reference section we list in brackets the page numbers on which that particular article is cited. We have found this to be an extremely useful tool in quickly locating information of interest. For the reader's convenience, we have provided an appendix including all the graphs on at most 6 nodes.

Distance in Graphs contains many theorems with proofs omitted in order to keep the book of manageable size while including a generous collection of interesting results. The reader is encouraged to look up the original references which we have been careful to include. By describing many unsolved problems, the text provides fertile background material for research investigations.

We have chosen to dedicate this book to two of our colleagues whose research on distance-related concepts have not only contributed a great deal to the excitement of graph theory, but has also inspired others to pursue study in this area. Frank Boesch obtained many interesting results on graph connectivity and popularized the use of circulants for studying vulnerability and reliability in graphs and networks. Peter Slater, who incidently is the academic grandson of F.H. via Steve Hedetniemi, introduced numerous centrality measures and obtained many interesting centrality results which are useful in both theoretical and applied settings.

We are indebted to many people for their assistance and consideration during the preparation of this book. We deeply appreciate the time and efforts of our colleagues Lowell Beineke, Norman Biggs, Marilyn Breen, Gary Chartrand, M.S. Krishnamoorthy, Mike Plummer, Lou Quintas, Allen Schwenk, Pete Slater, and Carsten Thomassen who read and commented on chapters of the book for us. We thank Addison-Wesley, in particular, our editor Allan Wylde for his patience and Jan Benes and Laura Likely for their help. This manuscript was prepared using the Addison-Wesley typesetting package MicroTEX and the figures were drawn with Autosketch. We sincerely thank both our typist and typesetter F. Buckley. He in turn asked us to thank John Buhrer at Arbortex for software installation assistance, and to express his sincere thanks to A-W electronic publishing consultants Laurie Petrycki and Mona Zeftel, who wrote most of the typesetting macros for this book.

F.B. thanks the School of Liberal Arts & Sciences of Baruch College (CUNY) who provided released time for this project; and the last portion of the book was completed during a semester sabbatical. F.H. thanks the Department of Computer Science and the Computer Research Laboratory of New Mexico State University. Together we would like to thank Stevens Institute of Technology, where we conceived this book in Spring 1984 while F.H. held the position of Visiting Research Professor of Electrical Engineering and Computer Science at that institution.

Finally, we thank our ladies Wai Mui Choy and Lucía Muñoz Hayakawa, respectively, for their patience, cheerfulness, and encouragement while we were often working away somewhere at a distance in graphs.

New York, NY   F.B.

Las Cruces, NM   F.H.

7 October 1989

Contents

PART TWO - NETWORKS AND ALGORITHMS

CHAPTER 1

Graphs

Graph theory is a branch of mathematics which has applications in many areas: anthropology, architecture, biology, chemistry, computer science, economics, environmental conservation, psychology, and telecommunications, to name a few. The list goes on and on. In a typical situation, a problem arises in a real-world subject area that can be modeled using graphs. Then existing theorems or algorithms are used or new ones are developed to solve the original problem. We describe the modeling process and present the basic concepts and terminology of graph theory with emphasis on the concept of distance in graphs. In addition, we describe a variety of graphs and useful graph operations.

1.1 GRAPHS AS MODELS

The concept of distance will be defined in the next section. In the present section we shall define graphs and see how they are used. Before doing so, however, we show in Figure 1.1 all eleven graphs with four nodes (of order 4).

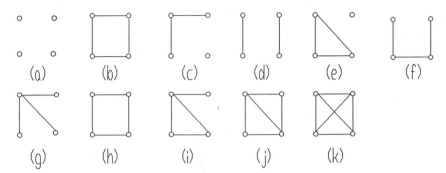

Figure 1.1   The eleven graphs with 4 nodes.

Later we will see that

1.   Every graph of order 4 is "isomorphic" to one of these.
2.   Graphs (a) - (e) are disconnected.
3.   (f) - (k) are connected.
4.   (k) is complete.
5.   (h) is a cycle.
6.   (f) is a path.
7.   (g) is a star.
8.   (f) and (g) are trees.

There is no choice at this point: We present here quite a list of formal definitions following the previous book [H14] by one of us. A *graph G* consists of a finite nonempty set $V = V(G)$ of p *nodes* together with a set E of q unordered pairs of distinct nodes of V. We say G has *order p* and *size q*. The pair $e = \{u, v\}$ of nodes in E is called an *edge* of G, and e is said to *join u* and v. We write $e = uv$ and say u and v are *adjacent nodes*. Adjacent nodes are said to be *neighbors*. Edge e is *incident* with each of its two nodes u and v. A graph with p nodes and q edges is called a (p, q)-*graph*.

To aid the intuition in using graphs, it is customary to represent a graph by means of a diagram and refer to it as the graph. Figure 1.2 shows a $(6, 8)$-graph in which s and t are adjacent and edge e is incident with node w. Nodes s and v are nonadjacent. A set of nodes is *independent* if the nodes are mutually nonadjacent. In Figure 1.2, $\{s, u, w\}$ is an independent set. The *degree* of node v is the number of edges incident with it and is denoted deg v. In Figure 1.2, deg $s = 2$ while deg $x = 4$.

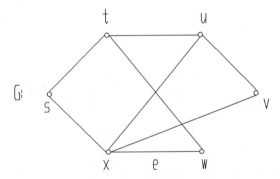

Figure 1.2   A graph to illustrate adjacency and incidence.

The first theorem of graph theory is due to Euler [E3].

Theorem 1.1   The sum of the degrees of the nodes of a graph is twice the number of edges,

$$\sum \deg v_i = 2q.$$

Proof   Since each edge e is incident with two nodes, e contributes 2 to the sum of the degrees of the nodes.   □

Corollary 1.1   In any graph, the number of nodes of odd degree is even.   □

The *degree sequence* of a graph is a list of the degrees of the nodes in nonincreasing order. The *minimum degree* among the nodes of a graph G is denoted $\delta(G)$ while the *maximum degree* $\Delta(G)$ is the largest such number. Thus, the graph in Figure 1.2 has degree sequence $(4, 3, 3, 2, 2, 2)$, so $\delta(G) = 2$ and $\Delta(G) = 4$. If all nodes have the same degree k, then G is called *regular*, or *k-regular*. We then speak of the *degree of* G and write $\deg G = k$. In Figure 1.1, graph (a) is 0-regular, (d) is 1-regular, (h) is 2-regular, and (k) is 3-regular. The 3-regular graphs are called *cubic* and have been studied extensively.

Isomorphic Graphs

A graph is completely determined by specifying its node and edge sets. However, two people may draw the graph differently. For example, in Figure 1.3 we show two drawings of the graph G having $V(G) = \{a, b, c, d, e\}$

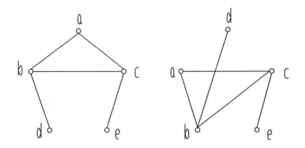

Figure 1.3   Two drawings of the same graph.

and $E(G) = \{ab, ac, bc, bd, ce\}$. Although the drawings appear different, they represent the same graph.

Two graphs G and H are *isomorphic* (written $G \cong H$ or sometimes $G = H$ and called *equal*) if there exists a one-to-one correspondence between their node sets which preserves adjacency. The graphs in Figure 1.3 are isomorphic. The three graphs G_1, G_2, and G_3 in Figure 1.4 are all isomorphic to one another. For example, G_1 and G_2 are isomorphic under the correspondence $v_i \longleftrightarrow u_i$. Can you find a labeling of G_3 to show it is isomorphic to G_1 and G_2?

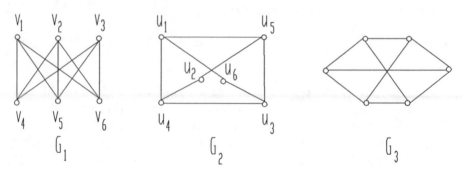

Figure 1.4   Three isomorphic graphs.

An *invariant* of a graph G is a number associated with G which has the same value for any graph isomorphic to G. We now have two simple invariants and one sequences of invariants we can use to distinguish a pair of nonisomorphic graphs:

1.   the number of nodes, p

2.   the number of edges, q

3.   the degree sequence.

We shall discuss isomorphic graphs further in the next section after several additional invariants are introduced.

A number of variations of graphs occur in applications. A *directed graph* or *digraph* D consists of a finite nonempty set V of nodes together with a collection A of ordered pairs of distinct nodes in V. The elements of A are called *arcs* or *directed edges*. A *symmetric pair* of arcs join two nodes u and v, one in each direction, i.e., arcs (u, v) and (v, u). An *oriented graph* is a digraph having no symmetric pair of arcs. In Figure 1.5 all four digraphs with three nodes and three arcs are shown; the last two are oriented graphs.

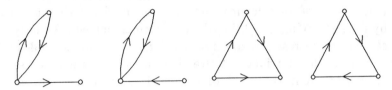

Figure 1.5   The digraphs with 3 nodes and 3 arcs.

We now illustrate how graphs are used as models with several examples.

Architecture

Foulds [F7] used graphs in the design and analysis of floor plans. Each room corresponds to a node, and two are adjacent if it is possible to go directly from one room to the other. In the building modeled by the graph of Figure 1.2, one can go directly from room t to room w but not directly from t to v. Node x has the largest degree and might represent a central meeting hall. The type of building often determines the properties desired in its floor plan and thus its resulting graph. For example. in designing a museum, one wants a structure where it is possible to walk through each hallway exactly once to see all the exhibits and leave the way one came in. A graph with this property is called *eulerian*. We will see later that such graphs are easy to characterize.

Computer Science

Graphs occur in very many situations in computer science. One such use is modeling computer networks, where each node represents a computer. An edge joins two nodes if there is a direct communication link between the corresponding computers.

Many algorithms such as the sorting algorithm "Heap Sort" use graphs in their development and analysis. Graphs are also used to analyze prefix and postfix operators used in computing (see McCracken [M3]).

A standard way of modeling a computer algorithm is to use a "flow-chart." The structure of a flowchart is a digraph. An algorithm whose flowchart gives the "simplest" digraph will often be most efficient.

Electrical Networks

Kirchhoff [K2] developed the theory of trees in 1847 to solve a system of simultaneous linear equations which give the current in each branch and around each circuit of an electrical network. He modeled the electrical network by its underlying graph in which electrical elements such as inductors, capacitors, or resistors are modeled by edges, and the nodes tell where two of these have a junction. He showed that it is not necessary to consider every cycle in the graph of an electrical network separately to solve the system of equations. Instead, he pointed out that it is sufficient to consider the "fundamental" system of independent cycles determined by any "spanning tree" of the graph.

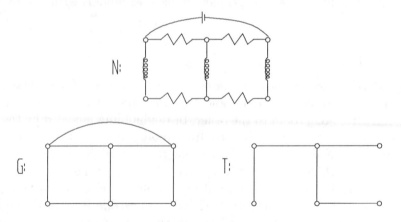

Figure 1.6   A network N, its underlying graph G, and a spanning tree.

Environmental Conservation

The nodes represent different species in a fixed geographical area. Regions called habitat patches in that geographic area are classified according to various qualities such as: near a stream, type of trees, hills, soil properties, vegetation, etc. An edge joins two nodes if the corresponding species share a common habitat patch. Chinn and Marcot [CM2] used graphs to analyze how proposed environmental management decisions may affect species interaction. Digraphs are often used to depict prey-predator relationships. The first such use of these competition digraphs was in [H9].

Games

In 1859 Sir William Hamilton invented a game that uses a regular solid dodecahedron (often used for paperweight desk calendars) with its 20 nodes labeled with the names of famous cities. The player is challenged to travel "Around the World" by finding a cycle which passes through each node exactly once and returns to the starting point. In graphical terms, the object of the game is to find a "spanning cycle" in the graph of the dodecahedron, shown in Figure 1.7.

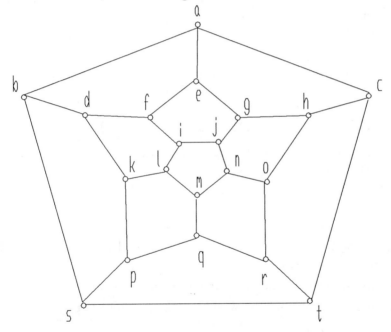

Figure 1.7   The graph of the dodecahedron.

A *spanning cycle* in a graph is a cycle that passes through each node exactly once. These are usually called *hamiltonian cycles*. They are very closely related to the Traveling Salesman's Problem and play a fundamental role in the theory of $\mathcal{NP}$-completeness which we discuss in Chapter 11.

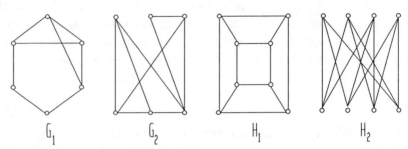

Figure 1.8 Two pairs of isomorphic graphs.

EXERCISES 1.1

1. Label the nodes in the graphs G_1 and G_2, and H_1 and H_2 of Figure 1.8 to show that $G_1 \cong G_2$ and $H_1 \cong H_2$.

2. Draw all graphs with three nodes.

3. Draw the graphs G and H with the following node and edge sets.

    a. $V(G) = \{a, b, c, d\}$, $E(G) = \{ab, ac, bd, cd\}$.

    b. $V(H) = \{e, f, g, h, i, j\}$, $E(H) = \{ef, ej, fi, fj, eg, ij\}$.

4. a. Draw all graphs having degree sequence $(3, 2, 2, 1, 1, 1)$.

    b. Why is $(5, 5, 4, 2, 2, 1, 1, 1)$ not the degree sequence of a graph?

5. Although $5 + 2 + 1 + 1 + 1 = 10$ (which is even), $(5, 2, 1, 1, 1)$ is not the degree sequence of a graph. Why?

6. The maximum number of edges in a graph of order p is $\binom{p}{2}$. [The words "Prove that" are understood.]

7. Let G be a (p, q)-graph all of whose nodes have degree k or $k + 1$. If G has $p_k > 0$ nodes of degree k and p_{k+1} nodes of degree $k + 1$, then $p_k = (k + 1)p - 2q$.

8. Draw all digraphs having four nodes and four arcs.

9. Find a "tour around the world" in the graph of Figure 1.7.

10. Draw all $(5, 4)$-graphs.

1.2   PATHS AND CONNECTEDNESS

One of the most basic properties any graph can enjoy is that of being connected. Informally, a graph is connected if it is all in one piece. In this section, we make this concept precise and examine several fundamental classes of connected graphs: paths, trees, and cycles.

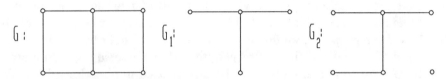

Figure 1.9 A graph and two subgraphs.

Subgraphs

A *subgraph* of G is a graph having all of its nodes and edges in G. It is a *spanning subgraph* if it contains all the nodes of G. If H is a subgraph of G, then G is a *supergraph* of H. For any set S of nodes in G, the *induced subgraph* $\langle S \rangle$ is the maximal subgraph with node set S. Thus two nodes of S are adjacent in $\langle S \rangle$ if and only if they are adjacent in G. In Figure 1.9, G_1 and G_2 are subgraphs of G. Here G_1 is an induced subgraph but G_2 is not; G_2 is a spanning subgraph but G_1 is not.

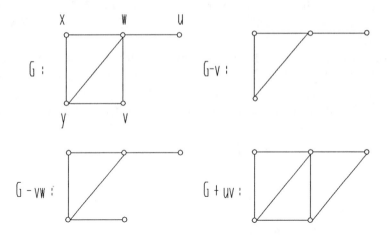

Figure 1.10 A graph minus a node; a graph plus or minus an edge.

The *removal of a node v* from a graph G results in that subgraph $G-v$ consisting of all nodes of G except v and all edges not incident with v. On the other hand, the *removal of an edge e* from G yields the spanning subgraph $G - e$ containing all edges of G except e. Thus $G - v$ and $G - e$ are the maximal subgraphs of G not containing v and e, respectively. If u and v are not adjacent in G, the addition of the edge uv results in the smallest supergraph of G containing the edge uv and is denoted $G + uv$. These concepts are illustrated in Figure 1.10.

There are certain graphs for which the result of deleting a node or edge or adding an edge is independent of the particular node or edge selected. If this is so for a graph G, we denote the result accordingly by $G - v$, $G - e$, or $G + e$. We shall see in future chapters that these highly symmetric graphs G play important roles in distance related applications. For now we mention that any cycle C_n is such a graph.

Walks and Paths

A *walk* in a graph G is an alternating sequence of nodes and edges $v_0, e_1, v_1, e_2, v_2, \ldots, v_{n-1}, e_n, v_n$ such that every $e_i = v_{i-1}v_i$ is an edge of G, $1 \leq i \leq n$. It is important to mention that the nodes need not be distinct and the same holds for the edges.

The walk *connects* v_0 and v_n and is sometimes called a v_0-v_n walk. This walk has *length* n, the number of occurrences of edges in it. A walk is a *trail* if all its edges are distinct and a *path* if all its nodes (and thus necessarily all its edges) are distinct. The walk is *closed* if $v_0 = v_n$ and is *open* otherwise. A closed walk is a *cycle* provided its n nodes are distinct and $n \geq 3$.

Since the edges in a walk are determined uniquely by writing its successive nodes, we usually do not list the edges. In the labeled graph G of Figure 1.11, a, b, e, b, c is a walk which is not a trail and a, b, e, d, b is a trail which is not a path; a, b, e, c is a path and b, d, e, b is a cycle.

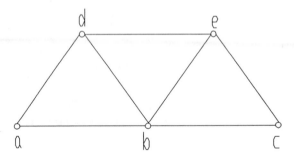

Figure 1.11   A graph to illustrate walks.

The *girth* of a graph G, denoted $g(G)$, is the length of a shortest cycle (if any) in G; the *circumference* $c(G)$ is the length of any longest cycle. Note that these terms are undefined if G has no cycles. The *distance* $d(u,v)$ between two nodes u and v in G is the minimum length of a path joining them if any; otherwise $d(u,v) = \infty$. A shortest u-v path is called a *u-v geodesic*. The *diameter* $d(G)$ of a connected graph G is the length of any longest geodesic. The graph G in Figure 1.11 has girth $g = 3$, circumference $c = 5$, and diameter $d = 2$.

Connected Graphs

A graph is *connected* if there is a path joining each pair of nodes. A *component* of a graph is a maximal connected subgraph. If a graph has only one component it is connected, otherwise it is *disconnected*. G and G_1 each have one component while G_2 has two.

Among important connected graphs of order p, a cycle is denoted by C_p and path by P_p. A *triangle* is a cycle C_3. The *complete graph* K_p has every pair of its p nodes adjacent. Thus K_p has $\binom{p}{2}$ edges and is regular of degree $p - 1$. Note that K_3 is the triangle C_3.

Isomorphic Graphs Revisted

In §1.1 we found several invariants helpful in distinguishing nonisomorphic graphs. We can now add to that list the following invariants:

1. the number of components

2. the diameter

3. the girth

4. the distances between pairs of nodes of a given degree.

Reconstruction

For a graph G with nodes $v_1, v_2, \ldots, v_p$, the graphs $G - v_i$ are often called the *node-deleted subgraphs* of G. Ulam [U1] suggested in the following conjecture that the collection of subgraphs $G - v_i$ of G completely determines G when $p \geq 3$.

The Kelly-Ulam Conjecture: Let G have p nodes v_i and H have p nodes u_i, with $p \geq 3$. If for each i, $G - v_i$ and $H - v_i$ are isomorphic, then the graphs G and H are isomorphic.

This conjecture is sometimes referred to as the reconstruction conjecture because of the following viewpoint of the problem [H13]. Draw each of the p unlabeled graphs $G - v_i$ on a 3×5 card thus obtaining the *deck* $D(G)$ of the graph G. A *legitimate deck* is one that can be obtained from some graph. Then the conjecture can be reformulated in terms of just one graph by asserting that any graph from which these subgraphs can be obtained by deleting one node at a time is isomorphic to G. Thus, one may try to prove that from any legitimate deck of cards, only one graph can be reconstructed. The conjecture has been proven true for several classes of graphs including regular graphs, disconnected graphs, and the class of graphs we shall now discuss — trees.

Trees

Perhaps the most important type of graph is a tree. This is so because of their applications to many different fields. In fact, an entire book is now being written on the theory of trees and their applications [HP4]. Furthermore, their simplicity makes it possible to investigate a conjecture for graphs by first studying it for trees. A graph is *acyclic* if it has no cycles. A *tree* is a connected acyclic graph. Any graph without cycles is a *forest*, thus the components of a forest are trees. There are 11 different trees with seven nodes, as shown in Figure 1.12.

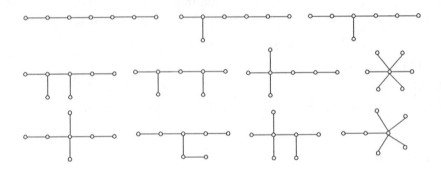

Figure 1.12   The eleven trees with seven nodes.

There are numerous ways of defining a tree as we shall now see.

Theorem 1.2   The following statements are equivalent.

1. G is a tree.
2.   Every two nodes of G are joined by a unique path.
3. G is connected and $p = q + 1$.
4. G is acyclic and $p = q + 1$.
5. G is acyclic and if any two nonadjacent nodes of G are joined by an edge e, then $G + e$ has exactly one cycle.

Proof $(1 \implies 2)$.   Since G is connected, every two nodes are joined by a path. Let P and P^* be two paths joining u and v in G, and let w be the first node of P (as we traverse P from u to v) such that w is on both P and P^* but its successor on P is not on P^*. If we let w^* be the next node on P which is also on P^*, then the segments of P and P^* which are between w and w^* together form a cycle in G. Thus if G is acyclic, there is at most one path joining any two nodes.

$(2 \implies 3)$.   Clearly G is connected. We prove $p = q + 1$ by induction. It is obvious for graphs of one or two nodes. Assume it is true for graphs with fewer than p nodes. Suppose G has p nodes ($p \geq 2$), q edges, and let v be a node of degree one (there must be such a node because of the uniqueness of paths, connectedness, and $p \geq 2$) in G. Then $G - v$ has $p - 1$ nodes, one less edge than G, and still satisfies property (2). By the inductive hypothesis, $G - v$ has order $p - 1 = (q - 1) + 1$. Thus the number of nodes in G is $p = q + 1$.

$(3 \implies 4)$.   Assume G has a cycle of length n. Then there are n nodes and n edges on the cycle, and for each of the $p - n$ nodes not on the cycle there is an incident edge on a geodesic to a node of the cycle. Each such edge is different, so $q \geq p$, which is a contradiction.

$(4 \implies 5)$.   Since G is acyclic, each component of G is a tree. If there are k components, then since each component has one more node than edge, $p = q + k$, so $k = 1$ and G is connected. Thus G is a tree and there is exactly one path connecting any two nodes of G. If we add an edge uv to G, that edge together with the unique path in G joining u and v forms a cycle, The cycle is unique because the path is unique.

$(5 \implies 1)$.   The graph G must be connected, for otherwise an edge e could be added joining two nodes in different components, and $G + e$ would be acyclic. Thus G is connected and acyclic, so G is a tree.   □

We need to specify a certain graph to eliminate trivial exceptions from theorems. The $(1, 0)$-graph is *trivial* following the terminology of group theory in which the "trivial group" has just one element.

Corollary 1.2   Every nontrivial tree has at least two endnodes.

Proof   Let P be a longest path in a nontrivial tree T and let u and v be endnodes of P. Since T is acyclic, u and v each have only one neighbor in P, and since P is a longest path they have no neighbors in $T - P$. Thus, there must be at least two nodes of degree one in a nontrivial tree.   □

Bipartite Graphs

A tree is a special type of a bipartite graph. A graph G is *bipartite* if its node set V can be partitioned into two subsets V_1 and V_2 such that every edge of G joins a node of V_1 with a node of V_2. For example, the graph of Figure 1.13a can be redrawn in the form of Figure 1.13b to display the fact that it is bipartite. If G contains every edge joining V_1 and V_2, then G is a *complete bipartite graph*. In this case, if V_1 and V_2 have m and n nodes, we write $G = K_{m,n}$. Obviously $K_{m,n}$ has mn edges. A *star* is a complete bipartite graph $K_{1,n}$. The *complete n-partite graph* $K(p_1, p_2, \ldots, p_n)$ has a node set V that can be partitioned into n parts $V_1, V_2, \ldots, V_n$ so that V_i has p_i nodes and two nodes are adjacent if and only if they are in distinct parts. Thus, a complete bipartite graph is a complete multipartite graph with just two parts.

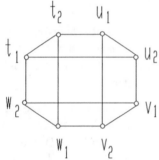 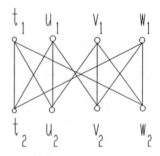

Figure 1.13   A bipartite graph.

It follows from a result of König [K4, p. 170] that a bipartite graph can have no odd cycles.

Theorem 1.3   A graph is bipartite if and only if all its cycles are even.

Proof   If G is bipartite, then its node set V can be partitioned into two sets V_1 and V_2 so that every edge of G joins a node of V_1 with a node of V_2. Thus, every cycle $v_1, v_2, \ldots, v_n, v_1$ in G necessarily has its oddly subscripted nodes in V_1, say, and the others in V_2, so that its length is even.

For the converse, we assume without loss of generality, that G is connected (for otherwise we can consider the components of G separately). Take any node $v_1 \in V$ and let V_1 consist of v_1 and all nodes at even distance from v_1, while $V_2 = V - V_1$. Since all the cycles of G are even, every edge of G joins a node of V_1 with a node of V_2. For suppose there is an edge uv joining two nodes of V_1. Then the union of geodesics from v_1 to v and from v_1 to u together with the edge uv contains an odd cycle, a contradiction.   □

Theorem 1.3 provides an easy method to determine whether a graph G is bipartite. Choose a node in each component of G and label it 1; then label its neighbors 2. Next label each unlabeled node adjacent to a node labeled 2 with the number 1, and continue alternating in this manner until every node is labeled. It is helpful to think of labels 1 and 2 as two colors. Graph G is bipartite if and only if the nodes of each color are independent.

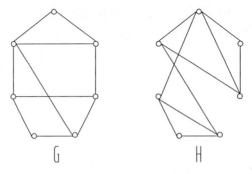

Figure 1.14   A pair of nonisomorphic graphs.

EXERCISES 1.2

1. Draw all graphs with five nodes. There are exactly 34.
2. Draw all graphs with degree sequence $5, 3, 2, 2, 2, 1, 1$. (Hint: There are just four.)
3. Every closed walk of odd length contains a cycle.
4. The graphs G and H in Figure 1.14 are not isomorphic.
5. For each $n \geq 3$, construct a cubic graph with $2n$ nodes having no triangles.
6. Prove or disprove:

    a. The union of any two different walks connecting two distinct nodes u, v contains a cycle.

    b. The union of any two different paths connecting u and v contains a cycle.

7. In a connected graph any two longest paths have a node in common.

8. A graph G is connected if and only if for any partition of V into two subsets V_1 and V_2, there is an edge of G joining a node of V_1 with a node of V_2.

9. If G has p nodes and $\delta(G) \geq (p-1)/2$, then G is connected.

10. Determine the maximum number of edges in a graph with p nodes and no even cycles.

11. Every tree is bipartite.

12. Every connected graph has a spanning tree.

13. Draw all trees with 8 nodes.

14. Determine the number of edges in $K(p_1, p_2, \ldots, p_n)$.

15. If G is a (p,q)-graph for which $q < p-1$, then G is disconnected.

16. Draw all cubic graphs with eight nodes.

17. Draw all 4-regular graphs with seven nodes.

18. If $\delta(G) \geq 2$ then G contains a cycle.

19. A tree with $p \geq 3$ has diameter 2 if and only if it is a star.

20. Prove or disprove:

    a. If G has a spanning star, then it has diameter 2.

    b. If G has diameter 2, then it has a spanning star.

1.3 CUTNODES AND BLOCKS

Some connected graphs can be disconnected by the removal of a single node called a cutnode. The distribution of such nodes is of considerable assistance in the recognition of the structure of connected graphs. Edges with the analogous cohesive property are known as bridges. The fragments of a graph held together by its cutnodes and bridges are called its blocks. We characterize these concepts in this section.

A *cutnode* of a graph is a node whose removal increases the number of components, and a *bridge* is such an edge. Thus if v is a cutnode of a connected graph G, then $G - v$ is disconnected. A *nonseparable* graph is connected, nontrivial, and has no cutnodes. A *block of a graph* is a maximal nonseparable subgraph.

In Figure 1.15, v is a cutnode while w is not; edge x is a bridge but y is not; and the four blocks of G are displayed. Each edge of a graph lies in exactly one of its blocks, as does each node that is not isolated or a cutnode. Furthermore, the edges of any cycle of G lie entirely within a single block.

Theorem 1.4   Let v be a node of a connected graph G. Then v is a cutnode of G if and only if there exist nodes u and w distinct from v such that v is on every u-w path.

Proof   If v is a cutnode in the connected graph G, then $G - v$ is disconnected. Let u and w be nodes in distinct components of $G - v$. There are no u-w paths in $G - v$, but there are u-w paths in G since G is connected. Thus every u-w path in G contains v.

Conversely, if v is on every path in G joining u and w, then there cannot be a path joining these nodes in $G - v$. Thus $G - v$ is disconnected, so v is a cutnode of G.   □

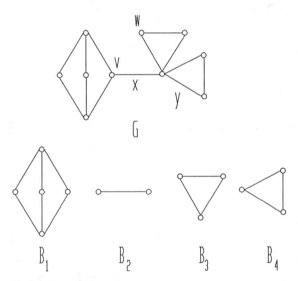

Figure 1.15   A graph and its blocks.

Theorem 1.5   Let e be an edge of a graph G. The following statements are equivalent.

1. e is a bridge.
2. e is not on any cycle of G.
3.   There exist nodes u and v of G such that the edge e of G is on every path joining u and v.   □

Any theorem, such as Theorem 1.5, that is stated without proof is followed by the end of proof symbol □ and the proof is generally quite routine and considered to be an exercise. On the other hand, when a specific reference is given, the proof is often too lengthy or too cumbersome for inclusion, and the reader may wish to consult the original source for further details.

There are many equivalent conditions for a graph to be a nonseparable, each of which may serve as the definition.

Theorem 1.6 Let G be a connected graph with at least three nodes. The following statements are equivalent.

1. G is nonseparable.
2. Every two nodes of G lie on a common cycle.
3. Every node and edge of G lie on a common cycle.
4. Every two edges of G lie on a common cycle.
5. Given two nodes and one edge of G, there is a path joining the nodes which contains the edge.
6. For every three distinct nodes of G, there is a path joining any two of them which contains the third.
7. For every three distinct nodes of G, there is a path joining any two of them which does not contain the third.

Proof $(1 \Longrightarrow 2)$ Let u and v be distinct nodes of G, and let U be the set of nodes different from u which lie on a cycle containing u. Since G has at least three nodes and no cutnodes, it has no bridges; therefore, every node adjacent to u is in U, so U is not empty.

Suppose v is not in U. Let w be a node in U for which the distance $d(w, v)$ is minimum. Let P_0 be a shortest w-v path, and let P_1 and P_2 be the two u-w paths of a cycle containing u and w (see Figure 1.16a). Since w is not a cutnode, there is a u-v path P' not containing w (see Figure 1.16b). Let w' be the node nearest u in P' which is also in P_0, and let u' be the last node of the u-w' subpath of P' in either P_1 or P_2. Without loss or generality, we assume u' is in P_1.

Let Q_1 be the u-w' path consisting of the u-u' subpath of P_1 and the u'-w' subpath of P'. Let Q_2 be the u-w' path consisting of P_2 followed by the w-w' subpath of P_0. Then Q_1 and Q_2 are distinct u-w' paths.

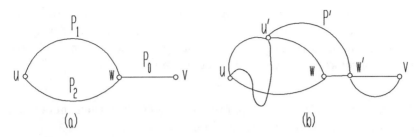

Figure 1.16   Paths in nonseparable graphs.

Together they form a cycle, so w' is in U. Since w' is on a shortest w-v path, $d(w', v) < d(w, v)$. This contradicts our choice of w, proving that u and v do lie on a cycle.

$(2 \Longrightarrow 3)$ Let u be a node and vw an edge of G. Let Z be a cycle containing u and v. A cycle Z' containing u and vw can be formed as follows. If w is on Z, then Z' consists of vw together with the v-w path of Z containing u. If w is not on Z, there is a w-u path P not containing v, since otherwise v would be a cutnode by Theorem 1.4. Let u' be the first node of P in Z. Then Z' consists of vw followed by the w-u' subpath of P and the u'-v path in Z containing u.

$(3 \Longrightarrow 4)$ The proof is analogous to the preceding one, and the details are omitted.

$(4 \Longrightarrow 5)$ Any two nodes of G are incident with one edge each, which lie on a cycle by (4). Hence any two nodes of G lie on a cycle , and we have (2), so also (3). Let u and v be distinct nodes and e an edge of G. By statement (3), there are cycles Z_1 containing u and e, and Z_2 containing v and e. If v is on Z_1 or u is on Z_2, there is clearly a path joining u and v containing e. Thus, we need only consider the case where v is not on Z_1 and u is not on Z_2. Begin with u and proceed along Z_1 until reaching the first node w of Z_2, then take the path on Z_2 joining w and v which contains e. This walk constitutes a path joining u and v that contains e.

$(5 \Longrightarrow 6)$ Let u, v, and w be distinct nodes of G, and let e be any edge incident with w. By (5), there is a path joining u and v which contains e, and hence must contain w.

$(6 \Longrightarrow 7)$ Let u, v, and w be distinct nodes of G. By statement (6), there is a u-w path P containing v. The u-v subpath of P does not contain w.

$(7 \Longrightarrow 1)$ By statement (7), for any two nodes u and v, no node lies on every u-v path. Hence, G must be nonseparable. $\square$

In a nontrivial connected graph G it is always possible to remove two nodes from G without disconnecting it. The following result of Chartrand, Kaugers, and Lick [CKL2] gives a condition for a nonseparable graph to have a nonseparable node-deleted subgraph.

Theorem 1.7   If G is nonseparable with $\delta(G) \geq 3$, then there is a node v such that $G - v$ is also nonseparable.   □

EXERCISES 1.3

1. What is the maximum number of cutnodes in a graph with p nodes?

2. A cubic graph has a cutnode if and only if it has a bridge.

3. The smallest number of nodes in a cubic graph with a bridge is 10.

4. A node v is a cutnode of a connected graph if and only if there exists a partition of the set of nodes $V - \{v\}$ into subsets U and W such that for any nodes $u \in U$ and $w \in W$, the node v is on every u-w path.

5. A node v of G is a cutnode if and only if there are nodes u and w adjacent to u such that v is on every u-w path.

6. Every nontrivial connected graph has at least two nodes that are not cutnodes.

7. A graph G is a tree if and only if G is connected and every edge of G is a bridge.

8. The following statements are equivalent:

    a. G is a forest.

    b. Every edge of G is a bridge.

    c. Every block of G is K_2.

    d. Every nonempty intersection of two connected subgraphs of G is connected.

9. If G is a nonseparable graph with $p \geq 3$, then given any two nodes and any edge, there is a path connecting the nodes which does not contain the edge.

10. A connected graph with at least two edges is nonseparable if and only if any two adjacent edges lie on a cycle.

11. Let $b(v)$ be the number of blocks to which v belongs in a connected graph G. Then the number of blocks of G is $b(G) = 1 + \sum[b(v) - 1]$.
(Harary [H7])

12. A graph G is *unicyclic* if it is connected and contains exactly one cycle. Prove: The following four statements are equivalent.

   a. G is unicyclic.

   b. G is connected and $p = q$.

   c. For some edge e of G, the graph $G - e$ is a tree.

   d. G is connected and the set of edges of G which are not bridges form a cycle.

13. If the degree of every node in G is even, then G has no bridges.

14. Every tree of order $p \geq 3$ contains a cutnode v such that every node except possibly one adjacent to v has degree 1.

1.4 GRAPH CLASSES AND GRAPH OPERATIONS

When a new concept is developed in graph theory, it is often first applied to particular classes of graphs. Afterwards, more general graphs are studied and theorems follow. In the previous sections, we encountered the paths, cycles, trees, and bipartite graphs. Many interesting graphs are obtained by combining pairs of graphs or operating on a single graph in some way. We now discuss a number of operations which are used to combine graphs to produce new graphs.

Complements

The *complement* $\overline{G}$ of a graph G has $V(G)$ as its node set, but two nodes are adjacent in $\overline{G}$ if and only if they are not adjacent in G. A graph and its complement are shown in Figure 1.17. The graphs $\overline{K}_p$ are called *totally disconnected*, and are regular of degree 0.

A well-known puzzle may be stated in the following form: Prove that at any party with six people, there are three mutual acquaintances or three mutual nonacquaintances. This can be rephrased in graphical terms. There is now an entire field of graph theory dealing with related problems. In fact, one of us has written a series of 17 papers (to date) on this subject, which is called Ramsey theory for graphs.

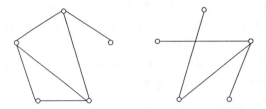

Figure 1.17   A graph and its complement.

Theorem 1.8   For any graph G of order 6, G or $\overline{G}$ contains a triangle.

Proof   Let v be a node of a graph G with six nodes. Since v is adjacent either in G or $\overline{G}$ to at least half of the five other nodes of G, we can assume without loss of generality that there are three nodes u_1, u_2, u_3 adjacent to v in G. If any two of these nodes are adjacent, then they are two nodes of a triangle whose third node is v. If no two of them are adjacent in G, then they are the nodes of a triangle in $\overline{G}$.   □

Theorem 1.9   If G is disconnected, then $\overline{G}$ is connected.   □

Self-Complementary Graphs

A *self-complementary* graph is isomorphic with its complement.

   Our first result about self-complementary graphs specifies their order.

Theorem 1.10   If G is self-complementary, then $p = 4n$ or $4n + 1$.   □

Theorem 1.11   If $d(G) \geq 3$, then $d(\overline{G}) \leq 3$.

Proof   Let x and y be any two nodes in $\overline{G}$. Since $d(G) \geq 3$, there exist nodes u and v at distance 3 in G. Hence uv is an edge in $\overline{G}$. Since u and v have no common neighbor in G, both x and y are each adjacent to u or v in $\overline{G}$. It follows that $d(x, y) \leq 3$ in $\overline{G}$, and hence $d(\overline{G}) \leq 3$.   □

   Harary and Robinson [HR1] derived Theorem 1.11 in order to give a simple proof of the following result first discovered independently by Ringel [R6] and Sachs [S3].

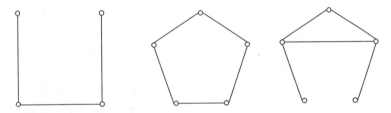

Figure 1.18   The smallest nontrivial self-complementary graphs.

Corollary 1.11   Every nontrivial self-complementary graph has diameter 2 or 3.   □

Operations on Graphs

It is rather convenient to be able to express the structure of a given graph in terms of smaller and simpler graphs. We discuss here only those operations that have been most useful in distance problems. Descriptions of others operations can be found in [H14]. Throughout this section, graphs G_1 and G_2 have disjoint node sets V_1 and V_2 and edge sets E_1 and E_2 respectively. Their *union* $G = G_1 \cup G_2$ has, as expected, $V = V_1 \cup V_2$ and $E = E_1 \cup E_2$. Their *join*, defined by Zykov [Z6], is denoted $G_1 + G_2$ and consists of $G_1 \cup G_2$ and all edges joining V_1 with V_2. In particular, $K_{m,n} = \overline{K}_m + \overline{K}_n$. These operations are illustrated in Figure 1.19 with $G_1 = K_2 = P_2$ and $G_2 = K_{1,2} = P_3$.

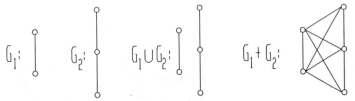

Figure 1.19   The union and join of two graphs.

For any connected graph G, we write nG for the graph with n components each isomorphic with G. Then every graph can be written in the form $\bigcup n_i G_i$ with G_i different from G_j for $i \neq j$. There are several operations on G_1 and G_2 whose set of nodes is the cartesian product $V_1 \times V_2$. These include the product and the composition [H8].

To define the *(cartesian) product* $G_1 \times G_2$ consider any two nodes $u = (u_1, u_2)$ and $v = (v_1, v_2)$ in $V = V_1 \times V_2$. Then u and v are adjacent

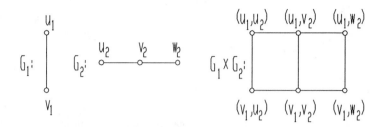

Figure 1.20   The product of two graphs.

in $G_1 \times G_2$ whenever $[u_1 = v_1$ and $u_2 v_2 \in E(G_2)]$ or $[u_2 = v_2$ and $u_1 v_1 \in E(G_1)]$. The cartesian product of $G_1 = P_2$ and $G_2 = P_3$ is shown in Figure 1.20.

The *corona* $G_1 \circ G_2$ was defined by Frucht and Harary [FH1] as the graph G obtained by taking one copy of G_1 of order p_1 and p_1 copies of G_2, and then joining the i'th node of G_1 to every node in the i'th copy of G_2. For the graphs $G_1 = K_2$ and $G_2 = P_3$, the two different coronas $G_1 \circ G_2$ and $G_2 \circ G_1$ are shown in Figure 1.21.

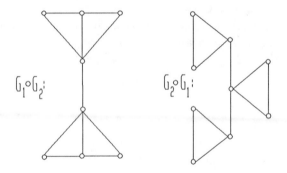

Figure 1.21   The two different coronas of two graphs.

If G_1 and G_2 are (p_1, q_1) and (p_2, q_2) graphs, respectively, then for each of the above operations one can calculate the number of nodes and edges in the resulting graph, as shown in Table 1.1.

An especially important class of graphs now known as hypercubes are most naturally expressed in terms of products. The *n-cube* Q_n is defined recursively by $Q_1 = K_2$ and $Q_n = K_2 \times Q_{n-1}$. Thus Q_n has 2^n nodes which may be labeled $a_1 a_2 \cdots a_n$, where each a_i is either 0 or 1. Two nodes of Q_n are adjacent if their binary sequences differ in exactly one place.

Table 1.1 BINARY OPERATIONS ON GRAPHS

Operation	Number of nodes	Number of edges
Union $G_1 \cup G_2$	$p_2 + p_2$	$q_1 + q_2$
Join $G_1 + G_2$	$p_1 + p_2$	$q_1 + q_2 + p_1 p_2$
Product $G_1 \times G_2$	$p_1 p_2$	$p_1 q_2 + p_2 q_1$
Corona $G_1 \circ G_2$	$p_1(1 + p_2)$	$q_1 + p_1 q_2 + p_1 p_2$

Figure 1.22 shows the 2-cube and the 3-cube, appropriately labeled. The importance of hypercubes lies in their usefulness in computer architecture.

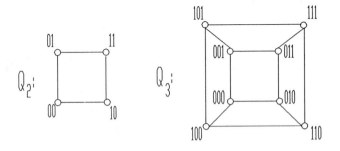

Figure 1.22   Two cubes.

An important graph formed by the join of other graphs is the wheel. For $n \geq 3$, the *wheel* $W_{1,n}$ is defined to be the graph $K_1 + C_n$. Buckley and Harary [BH4] defined the *generalized wheel* $W_{m,n}$ as the graph $\overline{K}_m + C_n$.

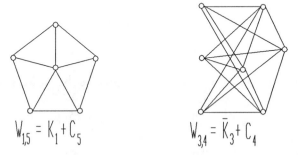

Figure 1.23   A wheel and a generalized wheel.

An operation used in recent work on distance introduced by Akiyama and Harary [AH2] is the sequential join. For three or more disjoint graphs

$G_1, G_2, \ldots, G_n$, the *sequential join*

$$G_1 + G_2 + \cdots + G_n$$

is the graph

$$(G_1 + G_2) \cup (G_2 + G_3) \cup \cdots \cup (G_{n-1} + G_n).$$

The graphs $K_1 + C_4 + K_1$ and $K_1 + K_1 + K_3 + K_1$ are shown in Figure 1.24.

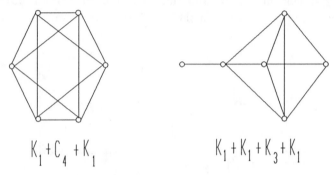

Figure 1.24   Two sequential joins.

Powers of a Graph

The *square* G^2 of a graph G introduced by Harary and Ross [HR2] has $V(G^2) = V(G)$ with u,v adjacent in G^2 whenever $d(u,v) \leq 2$ in G. The higher powers G^3, G^4, ... of G are defined similarly. Powers of graphs have been studied mostly in connection with hamiltonicity (Chapter 4) and chordal graphs (Chapter 2).

Clique Graphs

A *clique* of a graph is a maximal complete subgraph. The *clique graph* $K(G)$ of a given graph G has the cliques of G as its nodes and two nodes of $K(G)$ are adjacent if the corresponding cliques intersect. Not every graph is the clique graph of some graph. Roberts and Spencer [RS1] characterized clique graphs:

Theorem 1.12   A graph G is a clique graph if and only if it contains a family $\mathcal{F}$ of complete subgraphs whose union is G, such that whenever every pair of such complete graphs in some subfamily $\mathcal{F}'$ have a nonempty intersection, the intersection of all the members of $\mathcal{F}'$ is not empty.   □

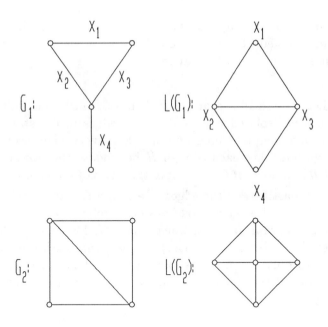

Figure 1.25   Graphs and their line graphs.

Line Graphs

Let graph G have at least one edge. The set of nodes of *line graph* of G, denoted $L(G)$ consists of the edges of G with two nodes of $L(G)$ adjacent whenever the corresponding edges of G are. Two examples of graphs and their line graphs are given in Figure 1.25. Note that in this figure $G_2 = L(G_1)$, so that $L(G_2) = L(L(G_1))$. We write $L^2(G) = L(L(G))$, and in general the *iterated line graph* $L^n(G) = L(L^{n-1}(G))$.

A graph G is *a line graph* if it is isomorphic to the line graph $L(H)$ of some graph H. For example $K_4 - e$ is a line graph; see Figure 1.25. On the other hand, we now verify that $K_{1,3}$ is not a line graph. Assume $K_{1,3} = L(H)$. Then H has four edges a,b,c,d since $K_{1,3}$ has four nodes. In H one of the edges, say a, is adjacent with the other three edges, while none of b,c,d are adjacent. Since a has only two endnodes, at least one pair of b,c,d must be adjacent to a at a single node, making that pair of edges adjacent to one another as well, a contradiction. So $K_{1,3}$ is not a line graph. By the same reasoning $K_{1,3}$ cannot be an induced subgraph or a line graph. The first characterization of line graphs is due to Krausz [K5].

Theorem 1.13   A graph G is a line graph if and only if the edges of G can be partitioned into complete subgraphs in such a way that no node lies in more than two of the subgraphs.

Proof   Let G be the line graph of H. Without loss of generality, we assume that H has no isolated nodes. Then the edges in the star at each node of H induce a complete subgraph of G and every edge lies in exactly one such subgraph. Since each edge of H belongs to the stars of exactly two nodes of H, no node of G is in more than two of the complete subgraphs.

Given a partition of the edges of a graph G into complete subgraphs $S_1, S_2, \ldots, S_n$ such that no node lies in more than two of the subgraphs, we construct a graph H whose line graph is G. The nodes of H correspond to the set S of subgraphs $S_1, S_2, \ldots, S_n$ together with the set U of nodes belonging to only one of the subgraphs S_i. Thus $S \cup U$ is the node set of H and two of these nodes are adjacent whenever they have nonempty intersection.   □

A triangle T of a graph G is called *odd* if there is a node of G adjacent to an odd number of its nodes, and is *even* otherwise. Another characterization of line graphs is given in the following theorem due to vanRooij and Wilf [RW1].

Theorem 1.14 G is a line graph if and only if

1. $K_{1,3}$ is not an induced subgraph of G, and
2. if $K_4 - e$ is an induced subgraph of G, then at least one of the two triangles in $K_4 - e$ is even.   □

Beineke [B3] displayed exactly those subgraphs which cannot occur as an induced subgraph of any line graph. This result can be proved by using Theorem 1.14.

Corollary 1.14   Graph G is a line graph if and only if none of the nine graphs of Figure 1.26 is an induced subgraph of G.

Proof   *(Outline).* Using Theorem 1.14, we see that $K_{1,3}$ is not an induced subgraph of a line graph G. Suppose $K_4 - e$ is an induced subgraph of G. Then to find other forbidden subgraphs, check possible adjacencies among nodes in G with nodes of $K_4 - e$ which would make both of its triangles

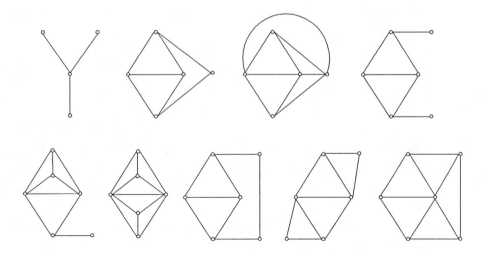

Figure 1.26 The nine forbidden induced subgraphs for line graphs.

odd, contradicting condition 2 of Theorem 1.14. For example, if some node v is adjacent to both of the nodes of degree two in $K_4 - e$ and no others, then both triangles are odd so G is not a line graph. In this case, we get the second graph of Figure 1.26 as a forbidden subgraph. If a node v is adjacent to all the nodes of $K_4 - e$, again both triangles are odd so G is not a line graph and we find the third forbidden subgraph. If nodes u and v are each adjacent to one of the nodes of degree 2 and no other nodes in $K_4 - e$, both triangles are odd so G is not a line graph and we find the fourth forbidden subgraph. Each of the other forbidden subgraphs is found in a similar manner using Theorem 1.14. □

EXERCISES 1.4

1. Draw the 4-cube Q_4.
2. Prove or disprove: If G_1 and G_2 are regular, then so is
   a. $G_1 + G_2$ b. $G_1 \times G_2$ c. $G_1 \circ G_2$.
3. Prove or disprove: If G_1 and G_2 are bipartite, then so is
   a. $G_1 + G_2$ b. $G_1 \times G_2$ c. $G_1 \circ G_2$.
4. Construct the following graphs.
   a. $K_3 + P_4$ b. $P_3 \times K_3$ c. $P_3 \circ K_{1,3}$.
5. Prove or disprove:
   a. $\overline{G_1 + G_2} = \overline{G_1} + \overline{G_2}$ b. $\overline{G_1 \times G_2} = \overline{G_1} \times \overline{G_2}$

6. For the graph $G = K_1 + K_1 + P_3 + K_1$, construct each of the following:
    a. $\overline{G}$ b. G^2 c. $L(G)$.

7. If $d(u,v) = m$ in G, what is $d(u,v)$ in the nth power G^n?

8. A graph G and its complement $\overline{G}$ are both connected if and only if no complete bipartite graph spans G or $\overline{G}$.

9. If v is a cutnode of G, then v is not a cutnode of $\overline{G}$.

                                                        (Harary [H5])

10. The square of every nontrivial connected graph is nonseparable.

11. Graph G is a tree if and only if G is not $K_3 \cup K_1$ or $K_3 \cup K_2$, $p = q+1$, and if any two nonadjacent nodes of G are joined by an edge e, then $G + e$ has exactly one cycle.

12. If G is a graph whose nodes have degrees d_i, then the number of edges in the line graph $L(G)$ is $\frac{1}{2} \sum d_i(d_i - 1)$.

13. Two regular graphs G and H of the same size whose line graphs also have the same size are of the same degree and have the same order.

14. If G is k-regular with p nodes, then the number of edges in the iterated line graph $L^n(G)$ is given by

$$(p/2^{n+1}) \prod_{j=1}^{n} (2^j k - 2^{j+1} + 2).$$

15. A graph is *semiregular* if each node has one of two possible degrees d_1 or d_2. If $L(G)$ is regular, then either G is regular or G is a semiregular bipartite graph in which all nodes in each partite set have the same degree.   (Ray-Chaudhuri [R2])

FURTHER RESULTS

Other graphs operations such as the conjunction of two graphs are developed in Harary and Wilcox [HW1] as boolean operations. For additional references to other graph operations, see Harary [H14].

CHAPTER 2

Centers

Facility location problems deal with the task of choosing a site subject to some criterion. For example, in determining where to locate an emergency facility such as a hospital or fire station, we would like to minimize the response time between the facility and the location of a possible emergency. In deciding the position for a service facility such as a post office, power station, or employment office, we want to minimize the total travel time for all people in the district. When constructing a railroad line, pipeline, or superhighway, we want to minimize the distance from the new structure to each of the communities to be served. Each of these situations deals with the concept of centrality. However, the type of center differs for each of the three examples mentioned. Centrality questions are now examined using graphs and distance concepts. We shall see that various kinds of center are useful in facility location problems.

2.1 THE CENTER AND ECCENTRICITY

Let G be a connected graph and let v be a node of G. The *eccentricity* $e(v)$ of v is the distance to a node farthest from v. Thus

$$e(v) = \max\{d(u,v) : u \in V\}.$$

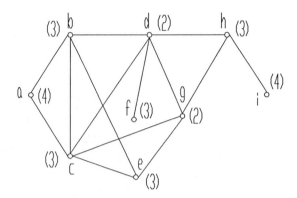

Figure 2.1   A graph and its eccentricities.

The *radius r(G)* is the minimum eccentricity of the nodes, whereas the *diameter d(G)* is the maximum eccentricity. Now v is a *central node* if $e(v) = r(G)$, and the *center C(G)* is the set of all central nodes. Thus, the center consists of all nodes having minimum eccentricity. Node v is a *peripheral node* if $e(v) = d(G)$, and the *periphery* is the set of all such nodes. For a node v, each node at distance $e(v)$ from v is an *eccentric node* for v. These concepts are illustrated in Figure 2.1 where the eccentricity of each node is shown in parenthesis. Graph G has radius 2, diameter 4 and central nodes d and g; nodes f and i are eccentric nodes for e.

   A basic result concerning centers is the classical theorem of Jordan [J2]. When $p(T) \geq 3$, let T' be the subtree of T obtained by removing all endnodes of T. A *caterpillar* is a tree for which the nodes that are not endnodes induce a path. In Figure 2.2, a tree T and its subtree T' is shown, and T' is a caterpillar.

Theorem 2.1   The center of a tree consists of either a single node or a pair of adjacent nodes.

Proof   The result is trivial for the trees K_1 and K_2. We show that any other tree T has the same center as the tree T'. Clearly, for each node v of T, only an endnode can be an eccentric node for v. Thus, the eccentricity of each node in T' will be exactly one less than the eccentricity of the same node in T. Hence the nodes with minimum eccentricity in T' are the same nodes of minimum eccentricity in T, that is T and T' have the same center. If the process of removing endnodes is repeated, we obtain successive trees having the same center as T. Since T is finite, we eventually obtain a subtree of T which is either K_1 or K_2. In either

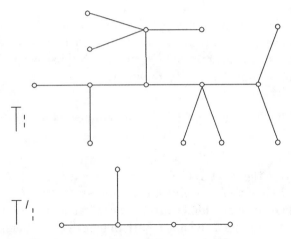

Figure 2.2   A tree T and its endnode deleted subtree T'.

case, the nodes in this ultimate tree constitute the center of T which thus consists of a single node or a pair of adjacent nodes.   □

A tree with one central node is called a *central tree* and one with two central nodes is called *bicentral*.

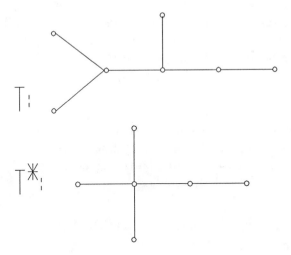

Figure 2.3   A central tree T and a bicentral tree T^*.

Recent work on centers has dealt with structure (what the subgraphs induced by the center look like), facility location (determining where to locate emergency and service facilities), embedding (determining when a supergraph G can be built around a graph H so that the nodes of H are precisely the central nodes of G), and central ratio (the ratio of the number of nodes in $C(G)$ the order of $V(G)$). The following generalization of Theorem 2.1 due to Harary and Norman [HN3] has been of fundamental importance in current centrality research.

Theorem 2.2   The center $C(G)$ of any connected graph G lies within a block of G.

Proof   Suppose the center $C(G)$ of a connected graph G lies in more than one block. Then G contains a cutnode v such that $G - v$ has components G_1 and G_2 each of which contains a central node of G. Let u be an eccentric node of v and let P be a u-v path of length $e(v)$. Then v contains no node from at least one of G_1 and G_2, say G_1. Let w be a central node of G_1 and let P' be a w-v geodesic in G. Then $e(w) \geq d(w,v) + d(v,u) \geq 1 + e(v)$. So w is not a central node, a contradiction. Thus all central nodes must lie in a single block.   □

The Centroid

A *branch* at a node v of a tree T is a maximal subtree containing v as an endnode. Thus, the number of branches at v is $\deg v$. The *weight* at a node v of T is the maximum number of edges in any branch at v. The weights at the nonendnodes of the tree in Figure 2.4 are indicated. Of course, the weight at each endnode is 13, the number of edges.

A node v is a *centroid node* of a tree T if v has minimum weight, and the *centroid* of T consists of all such nodes. Centroids have not been widely studied because, until recently, they were only defined for trees, and a theorem of Zelinka [Z4, Theorem 2.11] shows that their examination for trees would in some sense be redundant. Jordan also studied the centroid for trees and proved the following [J2].

Theorem 2.3   Every tree has a centroid consisting of either one node or two adjacent nodes.   □

Slater [S15] extended the concept of a centroid so that it is defined for all connected graphs. For a given pair of nodes u and v, let $c(u)$ be the number of nodes which are closer to u than to v, and let $c(v)$ be the number of nodes which are closer to v than to u. Let $f(u,v) = c(u) - c(v)$

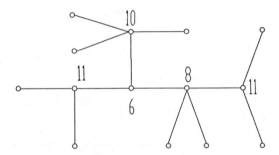

Figure 2.4   The weights at the nodes of a tree.

and let $g(u) = \sum_{v \in V - u} f(u, v)$. The *centroid of a graph* G is the set of all nodes for which $g(u)$ is maximum.

Structural Results

A graph is *planar* if it can be drawn in the plane with no crossing edges. Two graphs are *homeomorphic* if they can both be obtained from the same graph by a sequence of subdivisions of edges. For example, any two cycles are homeomorphic, and a graph homeomorphic to K_4 is displayed in Figure 2.5. Perhaps the most well-known structural result in all of graph theory is the celebrated theorem of Kuratowski [K8] which states that a graph is planar if and only if it has no subgraph homeomorphic to K_5 or $K_{3,3}$.

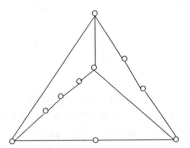

Figure 2.5   A homeomorph of K_4.

The *central subgraph* $\langle C(G) \rangle$ of a graph G is the subgraph induced by the center. Theorem 2.1 gives the structure of the central subgraph of a tree. A structural result for centers is of the following form: if G is a certain type of graph, e.g., *planar* or *unicyclic*, then the central subgraph

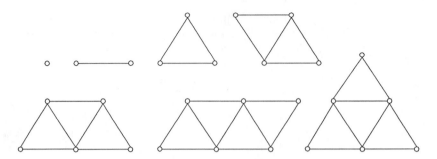

Figure 2.6   Central subgraphs of maximal outerplanar graphs.

of G must have a particular structure. Jordan's result asserts that if T is a tree, then $\langle C(T) \rangle$ is isomorphic to K_1 or K_2.

A graph is *outerplanar*, as defined by Chartrand and Harary [CH2], if it can be drawn in the plane with all nodes in the exterior boundary (face); it is *maximal outerplanar* if no edge can be added without destroying its outerplanar property. The next structural result on centers after Jordan (1869) was published by Proskurowski [P10] in 1980!

Theorem 2.4   If G is a maximal outerplanar graph, then its central subgraph $\langle C(G) \rangle$ is isomorphic to one of the seven graphs in Figure 2.6.

$\square$

A graph is *chordal* if every cycle of length greater than 3 has a chord. Every tree is (vacuously) a chordal graph and a maximal outerplanar graph is also chordal. The structure of the central subgraph has also been considered for chordal graphs by Laskar and Shier [LS1], and for other classes by Hedetniemi and Hedetniemi [HH3], and Hedetniemi, Hedetniemi, and Slater [HHS1].

Embedding and center size problems are discussed in the next section. Facility location problems involve algorithms for determining which nodes are central, so these problems will be discussed in Chapter 11 on Graph Algorithms.

Eccentricity

When an edge is added to a graph, the eccentricities of the nodes may be affected. A theorem of Ore [O3] tells precisely when the diameter decreases no matter where an edge is added. A graph G is *diameter-maximal* if for any edge $e \in E(\overline{G})$, $d(G + e) < d(G)$.

Theorem 2.5   A connected graph G is diameter-maximal if and only if

1. G has a unique pair of eccentric peripheral nodes u and v,
2. the set of nodes at each distance k from u induces a complete graph, and
3. every node at distance k is adjacent to every node at distance $k + 1$.

A disconnected graph G is diameter-maximal if and only if $G = K_m \cup K_n$.
□

Note that the structure of a connected diameter-maximal graph G can be described in terms of a sequential join. For some $d - 1$ positive integers a_i, G has the form

$$K_1 + K_{a_2} + K_{a_3} + \cdots + K_{a_d} + K_1.$$

A graph G is a *unique eccentric node graph* (*u.e.n.*) if each node in G has exactly one eccentric node. This concept was defined by Nandakumar an Parthasarathy [NP1] whose characterization of self-centered u.e.n. graphs will be given later, but now we display a simpler result.

Corollary 2.5   Every diameter-maximal graph with odd diameter is a u.e.n. graph.

Proof   Let G be a diameter-maximal graph with odd diameter d. By Theorem 2.5, G has two peripheral nodes u and v. For each node w in G with $d(u, w) \geq (d + 1)/2$, u is its unique eccentric node; v is the unique eccentric node for all other nodes in G. Thus G is a u.e.n. graph.   □

EXERCISES 2.1

1. For any connected graph G, the radius and diameter satisfy $r(G) \leq d(G) \leq 2r(G)$.
2. For any two positive integers m and n such that $m \leq n \leq 2m$ there is a connected graph G with $r(G) = m$ and $d(G) = n$.
3. A tree T has just one central node if and only if $d(T) = 2r(T)$.
4. Although a connected graph G satisfies $d(G) \leq 2r(G)$, the same is not true for digraphs. Construct a strongly connected digraph D with $r(D) = 2$ and $d(D) = 5$.

5. Construct all u.e.n. graphs of order 5 and 6.

6. Every u.e.n. graph G, $d(G) \le 2r(G) - 1$.

7. A tree T is a u.e.n. graph if and only if it has two central nodes and two peripheral nodes.   (Nandakumar and Parthasarathy [NP1])

8. The minimum number of edges in a nonseparable graph with diameter $d > 1$ is 4 if $p = 2d = 4$, and is $\lceil (pd - 2d - 1)/(d - 1) \rceil$ otherwise.
                                                            (Buckley [B21])

9. A tree T satisfies $C(\overline{T}) = V(T) - C(T)$ if and only if $d(T) = 3$.
                                                            (Buckley [B23])

10. If C is an even cycle in G and S is any subset of $V(G)$, then there exists a unicyclic subgraph of G with cycle C and center S.
                                            (Hedetniemi and Hedetniemi [HH3])

11. If G is regular with diameter 3, then $d(\overline{G}) = 2$.

12. If G is a connected chordal graph, then $\langle C(G) \rangle$ is connected and chordal.   (Laskar and Shier [LS1])

2.2 SELF-CENTERED GRAPHS

Some graphs G have the property that each node of G is a central node. A graph is *self-centered* if every node is in the center. Thus, in a self-centered graph G all nodes have the same eccentricity, so $r(G) = d(G)$. In this section, we discuss self-centered graphs as well as the problem of embedding a graph G in a supergraph H such that $\langle C(H) \rangle \cong G$. Buckley [B21] determined the extremal sizes of a connected self-centered graph having p nodes and radius r, anticipating a question of Bermond and Bollobás [BB1].

Theorem 2.6   Let $p \ge 5$ and $p \ge 2r > 2$. Then there exists a self-centered connected (p,q)-graph with radius r if and only if

$$\lceil (pr - 2r - 1)/(r - 1) \rceil \le q \le (p^2 - 4pr + 5p + 4r^2 - 6r)/2.$$

If $p = 2r = 4$ then q must be 4. $\square$

Corollary 2.6   If G is a self-centered (p,q)-graph with radius 2, then $q \ge 2p - 5$. $\square$

The *double star* $S_{m,n}$ is a tree of the form $\overline{K}_m + K_1 + K_1 + \overline{K}_n$. The graph $K_3(a,b,c)$ is formed by joining a, b, and c end edges to the 3 nodes

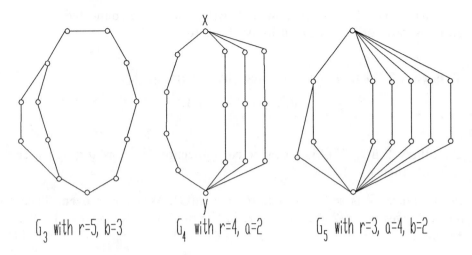

G_3 with r=5, b=3 G_4 with r=4, a=2 G_5 with r=3, a=4, b=2

Figure 2.7   Minimal self-centered graphs with given radius.

of K_3. Akiyama, Ando, and Avis [AAA1] used Corollary 2.6 to obtain
a structural characterization of the self-centered graphs shown in Figure
2.8 with p nodes and radius 2 of minimum size.

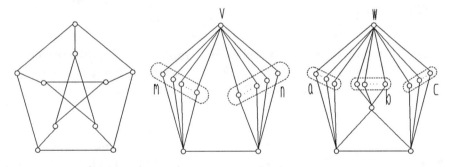

Figure 2.8   Minimum sized self-centered graphs with radius 2.

Theorem 2.7   If G is a self-centered graph of order p with radius 2 of
minimum size, then G is one of the following:

1.   The Petersen graph,

2.   The graph obtained from a double star $S_{m,n}$ by adding a new node v
     and joining v to every endnode of $S_{m,n}$,

3.   The graph obtained from any $K_3(a,b,c)$ by joining a new node w to
     each endnode of this $K_3(a,b,c)$.   □

Laskar and Shier [LS1] showed that the radius of a connected chordal graph is "almost" determined by its diameter.

Theorem 2.8   If G is a connected chordal graph, then

$$d/2 \leq r \leq d/2 + 1.$$ □

Corollary 2.8   If G is a connected, self-centered, chordal graph with radius r, then $r = 1$ or 2.

Proof   Since G is self-centered, $r(G) = d(G)$. When r is even, Theorem 2.8 gives $r \leq r/2 + 1$, so $r = 2$. In the same way, we get $r = 1$ when r is odd.   □

Nandakumar and Parthasarathy [NP1] obtained the following characterization.

Theorem 2.9   A u.e.n. graph G is self-centered if and only if each node of G is eccentric.

Proof   Let G be a self-centered u.e.n. graph. For an arbitrary node v, denote its eccentric node by v^*, so $d(v, v^*) = r(G)$. Then v is the eccentric node for v^*. Thus, each node of G is an eccentric node.

For the converse, we are given that each node v of a u.e.n. graph G is an eccentric node. We first show that $(v^*)^* = v$. Suppose not, and without loss of generality, assume u has least eccentricity among eccentric nodes v. Then $u = x^*$ for some node x. Note that $e(x) \leq e(u)$. If $e(x) = e(u)$ then $u^* = x$ so $(u^*)^* = x^* = u$, a contradiction. Thus assume $e(x) < e(u)$. Then $(x^*)^* = u^* \neq x$ and $e(x) < e(u)$ contrary to the choice of u. Thus $(v^*)^* = v$ for each v in G and $e(v) = e(v^*)$.

Suppose $r(G) < d(G)$. Then some pair of adjacent nodes w and v satisfy $e(w) < e(v)$. Their eccentric nodes satisfy
$e(w^*) = e(w) < e(v) = e(v^*)$, so w^* and v^* are distinct. Since w^* is unique for w, $d(w, v^*) < d(w, w^*)$ which gives

$$d(v, v^*) \leq d(v, w) + d(w, v^*) = 1 + d(w, v^*) < 1 + d(w, w^*).$$

So $e(v) = d(v, v^*) < 1 + d(w, w^*) = 1 + e(w)$. Since $e(w)$ and $e(v)$ are integers, $e(v) \leq e(w)$, a contradiction. Hence we must have $r(G) = d(G)$, that is, G is self-centered.   □

Embedding Problems

It is quite easy to embed, as an induced subgraph, any graph G in a super-graph H such that $\langle C(H) \rangle$ is isomorphic to G. Indeed, for any graph G, the sequential join $H = K_1 + K_1 + G + K_1 + K_1$ satisfies $\langle C(H) \rangle = G$. Thus, at most four additional nodes are required for the embedding. Let $f(G)$ be the minimum number of additional nodes required to embed G as the central subgraph of a supergraph H. Then $f(G) \leq 4$, and $f(G) = 0$ if and only if G is self-centered. Buckley, Miller, and Slater [BMS1] char-acterized trees that require i additional nodes, $0 \leq i \leq 4$ in such an embedding.

Theorem 2.10   For a tree T, $f(T) = 0$ only for $T = K_1$ or K_2; $f(T) \neq 1$ or 3. If $p \geq 3$, then $f(T) = 2$ if and only if all endnodes of T have the same eccentricity.   □

Clearly, $f(G) \neq 1$ for any graph since a diametral path in the super-graph H has two endnodes. Bielak [B7] and Chen [C8] found the first examples of graphs G for which $f(G) = 3$, and Liu [L6] constructed infi-nite classes of such graphs.

Besides simply embedding G as the central subgraph of a super-graph H, we may require that H have some prescribed property, for example, H is k-regular. Such questions were also studied in [BMS1].

EXERCISES 2.2

1. Draw a connected graph G for which $\langle C(G) \rangle$ is not connected. What is the smallest such graph?

2. Every connected self-centered graph satisfies the inequality
   $\Delta \leq p - 2r + 2.$ (Akiyama, Ando, and Avis [AAA2])

3. If G is self-centered with radius 2 then so is $\overline{G}$.

4. If G_1 and G_2 are self-centered with radii r_1 and r_2, then the cartesian product $G_1 \times G_2$ is self-centered with radius $r_1 + r_2$.

5. Construct a self-centered, self-complementary graph with eight nodes.

6. If G is a graph with $p \geq 9$ and $k \geq p+1$ then there exists a k-regular graph H for which $\langle C(H) \rangle = G$.

7. If G is a u.e.n. graph with radius 3, then G is either diameter-maximal or self-centered.   (Nandakumar and Parthasarathy [NP1])

8. Construct a unicyclic graph G which requires exactly three additional nodes in a supergraph H for which $\langle C(H) \rangle = G$.

9. If G is a self-centered with radius $r \geq 3$, then $\overline{G}$ is self-centered with radius 2.

2.3 THE MEDIAN

The center of a graph is important in applications involving emergency facilities where response time (distance) to each single location (node) in the region (graph) is critical. Suppose, instead, we consider a service facility such as a post office, shopping mall, bank, or power station. When deciding where to locate a post office, we want to minimize the average distance that a person serviced by the post office must travel. This is equivalent to minimizing the total distance traveled by all people within the district.

A new shopping mall should be situated so as to minimize the total distance to all of the customers in the region. This would make traveling to the mall as convenient as possible for their average customer.

Power stations, banks, and other general service facilities should be located so that the total distance is minimized. This second type of problem is the subject of the present section.

Let G be a connected graph. The *status s(v)* of a node v in G is the sum of the distances from v to each other node in G. This concept was introduced by Harary [H6]. The *median M(G)* of a graph G is the set of nodes with minimum status. The *minimum status ms(G)* of a graph G is the value of the minimum status; the *total status ts(G)* is the sum of all the status values. These concepts are illustrated in Figure 2.9. The number near each node is its status. The minimum status of G is 8, the total status is 70, and the median consists of nodes b, d, and e.

The following result of Zelinka [Z4] showed that studying the centroid of a tree is equivalent to studying the median of a tree.

Theorem 2.11   Node v is a centroid node of a tree T if and only if it is a median node.   □

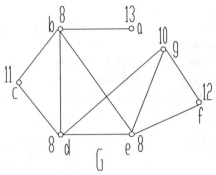

Figure 2.9   A graph and its statuses.

Thus Theorem 2.3 and Theorem 2.11 together imply that the median of a tree consists of either a single node or a pair of adjacent nodes. This also follows from the next more general result of Truszczyński [T10].

Theorem 2.12   The median $M(G)$ of any connected graph G lies within a block of G. $\square$

Entringer, Jackson, and Snyder [EJS1] found the range of $s(v)$ in a connected (p,q)-graph.

Theorem 2.13   For each node v of a connected (p,q)-graph G,

$$p - 1 \le s(v) \le (p-1)(p+2)/2 - q$$

and these bounds can be achieved for each q, $p - 1 \le q \le \binom{p}{2}$.

Proof   The easy lower bound is achieved by any (p,q)-graph with some node having degree $p-1$. We use induction on q to prove the upper bound holds. Since G is connected, begin with $q = p - 1$, so G is a tree. For any node v, let d_i be the number of nodes at distance i from v. Then

$$s(v) = \sum id_i \quad \text{and} \quad \sum d_i = p - 1.$$

Note that if $d_i = 0$, then $d_{i+1} = 0$. Thus the sum $\sum id_i$ is maximum when $d_i = 1$ for each i, so that

$$s(v) = \sum_{i=1}^{p-1} id_i = \frac{p(p-1)}{2} = \frac{(p-1)(p+2)}{2} - (p-1).$$

By the inductive hypothesis, the upper bound holds for any connected (p,q)-graph. Let v be a node in a connected $(p,q+1)$-graph G. Then G is not a tree, so it contains a cycle. Let u be a node in a cycle C such that $d(u,v)$ is minimum for such nodes (u is v if v is on a cycle). Let w be a node on C adjacent to u and consider $G - uw$. This graph is a connected (p,q)-graph. Because of the choice of u as a closest node to v in a cycle, $d(v,w)$ is greater in $G - uw$ than in G. By the inductive hypothesis,

$$s(v) \leq -1 + \frac{(p-1)(p+2)}{2} - q = \frac{(p-1)(p+2)}{2} - (q+1),$$

so the upper bound holds.

To show that the upper bound can be achieved for each value of q, $p - 1 \leq q \leq \binom{p}{2}$, let t be the largest integer for which $n = q - p + 1 - t(t-3)/2$ is nonnegative. Let G be the sequential join

$$K_1 + K_1 + \cdots + K_1 + K_n + K_{t-n},$$

where there are $(p-t)K_1$'s. This graph has p nodes and

$$(p - t - 1) + n + \binom{n}{2} + n(t-n) + \binom{t-n}{2}$$

edges. After simplifying and substituting for n, we find G is a (p,q)-graph. The node v of degree 1 in G has status

$$s(v) = \sum_{i=1}^{p-t-1} i + (p-t)n + (p-t+1)(t-n)$$

$$= \frac{(p-1)(p+2)}{2} - q. \qquad \square$$

In the same paper [EJS1], the following result is proven concerning the total status of a graph and its complement.

Theorem 2.14   For any graph G, $ts(G) + ts(\overline{G}) \geq 3p(p-1)$ and this bound is best possible for $p \geq 5$. $\qquad \square$

Self-Median Graphs

Analogous to self-centered graphs are the *self-median* graphs in which all nodes have the same status. Sabidussi [S2] noted that it is easy to

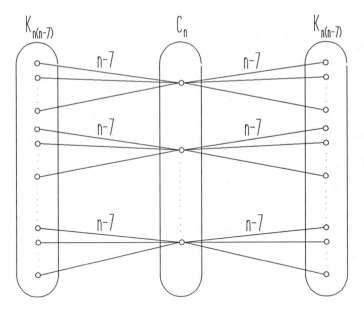

Figure 2.10   A self-median graph.

construct nonregular self-median graphs with radius r for any odd $r \geq 3$. His graphs have relatively few edges and have degree set $\{2, 3, 4\}$. If there is no restriction on the radius, we can construct nonregular self-median graphs with $\Delta - \delta$ arbitrarily large. Begin with C_n and two disjoint copies of $K_{n(n-7)}$. Then join each node of C_n to $n - 7$ nodes in each $K_{n(n-7)}$ so that each node in each $K_{n(n-7)}$ is adjacent to precisely one node of the cycle. See Figure 2.10.

A highly symmetric class of regular graphs, called distance-degree regular graphs, were defined and examined by Bloom, Kennedy, and Quintas [BKQ1]. These graphs will be discussed in Chapter 9 where we will see that in addition to their symmetry properties, such graphs are self-centered and self-median.

Embeddings

The *median subgraph* of a graph G is the induced subgraph $\langle M(G) \rangle$. Slater [S14] showed how to embed any graph G in a supergraph H so that $\langle M(H) \rangle = G$. If G is disconnected, then all nodes in G have the same status (∞), so G is self-median and no embedding is required. Thus, we may restrict attention to connected graphs. Miller [M7] later simplified Slater's construction by producing for any graph G with p nodes a supergraph H with at most $2p$ nodes so that $\langle M(H) \rangle = G$.

Theorem 2.15   Every graph G has a supergraph H whose median subgraph is isomorphic to G.

Proof   Let $V(G) = \{v_1 v_2 \cdots v_p\}$. Then form H as follows: add p new nodes $v'_1 v'_2 \cdots v'_p$ then join v'_i to v_i and to all nodes of G not adjacent to v_i. It is easy to check that in H, $s(v_i) = 3p - 2$ while $s(v'_i) = 3p - 2 + d_i$, where d_i is the degree of v_i in G. Since $d_i \geq 1$, $\langle M(H) \rangle = G$.   □

EXERCISES 2.3

1. Determine the status of each node in the following graphs:
    a. K_p b. P_n c. $K_{m,n}$ d. $W_{m,n}$.
    Determine the total status for each graph. Which of these graphs are self-median?

2. The graph G in Figure 2.10 has two types of nodes — those from C_n and those from some $K_{n(n-7)}$. Determine the status for each type of node to verify that G is self-median.

3. If u and v are adjacent nodes in a connected graph G, and x, y are the numbers of nodes closer to u than v, or to v than u, respectively, then $s(u) - s(v) = y - x$.

4. Every cube Q_n is self-median.

5. Construct graphs to illustrate each of the following:

    a. The median subgraph may be disconnected.

    b. A shortest path joining a minimum status node and a maximum status node need not contain a central node.

    c. There exists two different graphs of order 5 which have the same status list for their nodes.

6. An edge e of a connected graph G is a bridge if and only if each node of G has smaller status in G than in $G - e$.
                                   (Entringer, Jackson, and Snyder [EJS1])

7. Let G be a graph formed by a finite number of applications of the following operation. Add another C_3 and identify one of its edges with an edge already in the graph. Then the median subgraph of G is isomorphic to K_1, K_2, or K_3.   (Slater [S14])

8. Construct a tree with disjoint center and centroid, each having two nodes. What is the smallest such tree?

2.4   CENTRAL PATHS

If a superhighway is to be built connecting two major metropolitan areas so that it has a total of ten exits serving the towns in between, along what path should the highway be built and where should the exits be located to be most convenient to the largest number of people? If the towns all have the same political clout, the highway will be designed to minimize the maximum distance from the various towns to the closest exit. (There are other considerations such as cost and terrain which will be considered when we study networks in Chapter 11.)

Many years ago, the New York City Transit Authority had a New Routes Program which was designed to install a new subway line to service certain areas of Queens, one of the city's boroughs. If the plan called for a subway line with six stations to be built, where should the line and stations be located to best serve the people? In this case, there will be thousands of individuals using the new subway each day, and we want to minimize the sum of the distances that the people travel to the station closest to their home. (Since the New York fiscal crisis of the 1970's, the program has shrunk to the point where the line will have two stations — Roosevelt Island and Long Island City; the scheduled completion date was Fall 1985, and after numerous postponements the opening was last scheduled for October 1989.)

There are other situations such as the installation of natural gas pipelines or pipelines for irrigation, where one may want to find a path that all nodes in a graph are "close" to. In this section, we discuss several concepts which involve minimizing the distance to a path in a graph.

The Path Center

Let G be a graph, and let W be a subgraph of G. For any node v in G, the *distance $d(v, W)$ from v to W* is the minimum distance from v to a node in W. The *eccentricity of W*, $e(W)$, is the distance to a node farthest from W. Thus $e(W) = \max d(v, W)$ for v in G. We restrict our attention to the situation where W is a path in G. More general cases will be considered in the next section. A path P is a *path center* of G if P has minimum eccentricity and has minimum length among such paths. For the tree in Figure 2.11, paths *gfdik* and *abcdf* have eccentricity 3 and 2, respectively. The central path is *cd* with eccentricity 2.

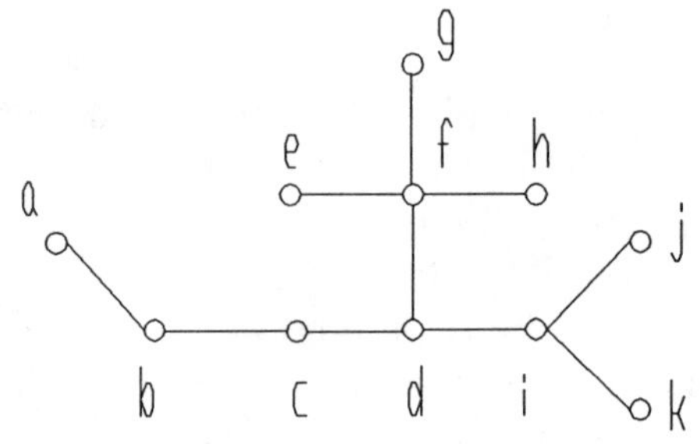

Figure 2.11   A tree to illustrate central paths.

The concept of path center was developed independently by Slater [S16] and by Cockayne, Hedetniemi, and Hedetniemi [CHH1]. They observed that the algorithm for finding the center of a tree is easily adapted to find the path center of a tree:

(1) If the tree is a path, stop. This is the path center. Otherwise,

(2) Delete all endnodes and go to (1).

A simple consequence is the following result.

Theorem 2.16   The path center of a tree T is unique and it contains the center $C(T)$.   □

It is easy to see that a similar statement does not extend from trees to graphs. For example, let G be any graph that is the join of two graphs G_1 and G_2 with p_1 and p_2 nodes ($K_{m,n}$ is such a graph), so $G = G_1 + G_2$. If neither G_1 nor G_2 is complete, then G has at least $p_1 p_2$ path centers — simply choose a path of length two with one node from G_1 and the other from G_2.

The Path Centroid

The *weight* $b(W)$ of a set W of nodes in a graph G is the number of nodes in the largest component of $G - W$. The *path centroid* of a graph G is a path with minimum weight having minimum length among such paths. The path centroid of the tree in Figure 2.11 is *fd* with weight 3. Slater [S19] showed that the analogous result to Theorem 2.16 holds for path centroids.

Theorem 2.17   The path centroid of a tree T is unique and it contains the centroid of T.

Proof   Let P be a path centroid and let k be the weight of a centroid node in tree T. By Theorem 2.3, the centroid consists of either one node or a pair of adjacent nodes. Assume P does not contain the centroid. Then P is a subgraph of one of the branches at a centroid node and $b(P) \geq k+1$. This is a contradiction since the subgraph induced by the centroid is a path with smaller branch weight. Thus, any path centroid contains the centroid.

Let W be the centroid of T, and let k be the weight of $w \in W$. Suppose $\langle V(T) - W \rangle$ contains three or more components with k nodes. Since a path P through W contains nodes from at most two of these components, $b(P) \geq k$. In this case, the shortest path is $\langle W \rangle$. If $\langle V(T) - W \rangle$ contains just two components with k nodes, let u and v be the nodes from those components which are adjacent to a node in W. With u and v in the path centroid P, $b(P) < k$, whereas if either u or v is not in P, then $b(P) \geq k$. Thus both u and v are in the path centroid. Note that u and v each have minimum weight among nodes adjacent to the centroid.

Suppose a path $P = v_1 v_2 \cdots v_n$ is known to be in the path centroid of T ($n \geq 2$). Let X_i be the set of nodes in $V(T) - V(P)$ which are adjacent to v_i. If $u \in X_1$ (or X_n) and $b(u) < b(v)$ for all $v \in \bigcup X_i - u$, then $P' = u v_1 v_2 \cdots v_n$ has $b(P') < b(P)$, so u is in the path centroid (if $u \in X_n$ then $P' = v_1 v_2 \cdots v_n u$). If $u_1 \in X_1$ and $u_n \in X_n$ such that $b(u_1) = b(u_n) < b(v)$ for all $v \in \bigcup X_i - \{u_1, u_n\}$, then $P'' = u_1 v_1 v_2 \cdots v_n u_n$ and both u_1 and u_n are in the path centroid. In all other cases, P cannot be extended to a path with smaller weight. We have thus described an algorithm to generate the unique path centroid of T. $\square$

Cores and Pits

The *status* $s(P)$ *of a path* P in graph G is the sum of the distances $d(v, V(P))$ for all $v \in V(G)$. A path with minimum status is a *core* or *path median* of G. The graph of Figure 2.8 has three cores — *abcdfe*, *abcdfg*, and *abcdfh* — each with status 7. Note that unlike the path center and path centroid, the core of a nontrivial tree will necessarily contain two endnodes of the tree. For each v in G, let $t(v)$ be the maximum difference between $s(v)$ and $s(P)$ where P is a nontrivial path with v as an endnode. A node v in G for which $t(v)$ is minimum is called a *pit node* and the set of such nodes is the *pit* of G. The pit of the tree in Figure 2.8 is $\{d\}$. Note

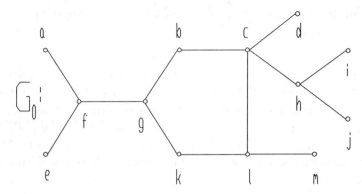

Figure 2.12 A graph to exercise centrally with.

that the center is $\{c,d\}$. It is easy to construct graphs with disjoint pit and center. Slater [S16] proved the following result about cores and pits of trees.

Theorem 2.18 Every tree T has a pit consisting of either one node or two adjacent nodes, and each core of T contains all its pits.   □

EXERCISES 2.4

1. Find all path centers, path centroids, and cores of the graph G_0 in Figure 2.12. Note that path centers and path centroids are unique for trees but not for graphs in general.
2. Determine the branch weight for each node in G_0.
3. Find the center, median, centroid, and pit for G_0.
4. Construct a tree with disjoint center and pit. What is the smallest such tree?
5. Construct a tree with disjoint core and path centroid. What is the smallest such tree?
6. It is not true that the path center of any *graph* G contains its center.

2.5   OTHER GENERALIZED CENTERS

A major new set of retirement communities are being planned for a state in the southwest and it is decided that there should be three firehouses to protect the houses within these communities. Where should the county locate the firehouses? Since they are emergency facilities, the firehouses should be situated to minimize the response time to the farthest point being protected.

Just recently, a major pizza company began opening stores throughout New York City. If they plan to open 15 stores, where should they be located? Such stores are considered general service facilities. Thus we want to situate the stores so that the sum of the distances from each customer to the closest store is minimized.

The central paths of the previous section are one type of generalized center. They are special cases of more general classes of problems — n-centers, n-medians, and n-centroids. When locating the set S of 3 firehouses to protect the retirement communities, this is an n-center problem (in reality, the county may already have two firehouses but wants to determine where to position a third one). A path center is an n-center for which the n nodes in S form a path. The problem of where to locate the pizza stores is an n-median problem. Here, we discuss n-centers, n-medians, the cutting center, the path centrix, and several other generalized centers.

The Cutting Center

This topic was introduced by Harary and Ostrand [HO1]. The *cutting number* $c(v)$ of a node v in a connected graph G is the number of pairs of nodes $\{u, w\}$ such that u and w are in different components of $G - v$. The *cutting center* $CC(G)$ of a graph G is the set of all nodes with maximum $c(v)$; a node in $CC(G)$ is called a *cutting center node*. Clearly, $c(v) > 0$ if and only if v is a cutnode. A graph G with its cutting numbers is displayed in Figure 2.13. The cutting center of G is $\{g\}$. Cutting centers have been studied mainly for trees, where every nonendnode has positive cutting number.

Harary and Ostrand [HO1] described the positioning of cutting center nodes for trees.

Theorem 2.19 The cutting center of a tree T is contained entirely in one path of the tree. $\quad\square$

It has also been shown that there are trees with an arbitrarily large cutting center as well as trees with two cutting center nodes which are arbitrarily far apart from one another.

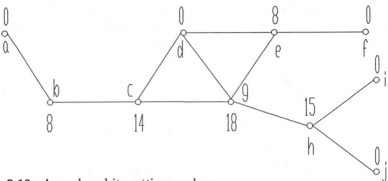

Figure 2.13   A graph and its cutting numbers.

Multicenters and Multimedians

For a graph G of order p, a set $S \subset V(G)$ is an *n-center* if $|S| = n$ and its eccentricity $e(S)$ is minimum among all n-subsets. Thus, an n-center is a set S of n nodes such that every other node is close to S, i.e., to at least one of the nodes in S. For a given n, a graph G may have several n-centers. For example, the graph G in Figure 2.14 has four 2-centers, which are $\{b,g\}$, $\{b,i\}$, $\{c,g\}$, and $\{c,i\}$, each having eccentricity 2.

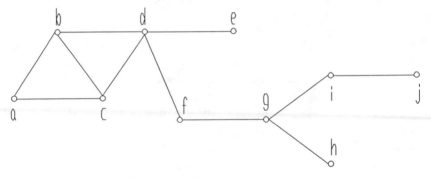

Figure 2.14   A graph with four 2-centers.

A simple result which produces a lower bound for the eccentricity of an n-center of a graph is the following:

Theorem 2.20   The eccentricity of an n-center of the path P_k is: $\lceil (k - n)/2n \rceil$.

Proof Let $P_k = v_1 v_2 \cdots v_k$ and suppose the eccentricity of an n-center is m. By being greedy, we may form an n-center as follows: choose v_{m+1}, v_{3m+2}, v_{5m+3}, $v_{7m+4}, \ldots, v_{(2j-1)m+j}$ while the subscripts are less than k. Then choose any other $n - j$ nodes. When k is as large as possible for a given m and n, we will choose $v_{(2n-1)m+n}$ which will be at distance m from v_k. In this extreme case, each node u of P_k has a unique node v in the n-center such that $d(u,v) \leq m$. There are $2m$ nodes on the path between successive n-center nodes, so

$$k \leq m + (n-1)2m + n + m = 2mn + n,$$

which gives $m \geq (k-n)/2n$. Since m is an integer, we get: $m = \lceil (k-n)/2n \rceil$.   □

Corollary 2.20   If G has diameter k, then the eccentricity of an n-center is at least $\lceil (k-n)/2n \rceil$.   □

A subset S of $V(G)$ is an *n-median* of G if $|S| = n$ and its status, $s(S) = \sum d(v, S)$, is minimum among all n subsets of $V(G)$. For a given n, a graph may have several n-medians. The graph G in Figure 2.14 has four 2-medians: $\{b, g\}$, $\{c, g\}$, $\{d, g\}$, and $\{d, i\}$, each with status 10.

The k-Centrum

Slater [S13] unified the concepts of the center and median of a graph by defining the k-centrum. For a node v, define its *k-centricity* $e_k(v)$ as the sum of the distances to the k nodes farthest from v. The *k-centrum* $C(G; k)$ of a graph G is the set of nodes for which $e_k(v)$ is minimum. Note that $C(G; 1)$ is the center of G and $C(G; p)$ is the median. The graph of Figure 2.15 has center $\{b, c, g, i\}$ and 3-centrum $\{c\}$.

Recently, Reid [R4] introduced a related but distinct concept, which he called the k-ball branch weight centroid for trees. For a nonnegative integer k and a node v in a tree, the k-ball branch weight $b(v; k)$ of v is the number of nodes in a largest subtree of T all of whose nodes are at distance at least $k + 1$ from v. The *k-ball branch weight centroid* $W(T; k)$ of a tree T is the set of all nodes v for which $b(v; k)$ is minimum. Reid examined structural properties of the subgraph spanned by $W(T; k)$.

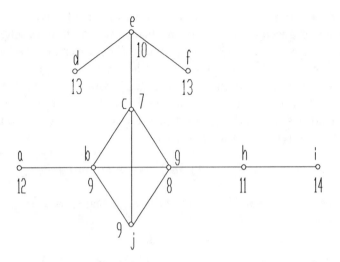

Figure 2.15   A graph and its 3-centricities.

The Accretion Center

An ordered n-tuple $(v_1, v_2, \ldots, v_n)$ is a *sequential labeling* of a connected graph G if $V(G) = \{v_1, v_2, \ldots, v_n\}$ and the subgraph induced by $\{v_1, v_2, \ldots, v_j\}$ is connected for each j, $1 \le j \le n$. For example, $(c, e, g, f, b, h, d, i, j, a)$ is a sequential labeling of the graph in Figure 2.15. For each node v, let $\sigma(v)$ be the number of sequential labelings of G which use v as the first node in the n-tuple. Vertex v is an *accretion center node* if $\sigma(v)$ is maximum among the nodes in G; the *accretion center* is the set of such nodes. Let uw be a bridge in in G, and let U and W be the components of $G - \{u, w\}$ containing u and w, respectively. Slater [S17] showed that the value of $\sigma(u)$ in G can be determined from the values of $\sigma(u)$ in U, $\sigma(w)$ in W, and the number of nodes in U and W, and is independent of the structure of the components U and W. Let n_u and n_w be the number of nodes in U and in W. For a node v in an induced subgraph H of G, let $\sigma(v; H)$ be the number of sequential labelings in H which use v as the first node.

Theorem 2.21   Suppose that G is a graph with bridge uw, $|V(U)| = n_u$, $|V(W)| = n_w$, and $|V(G)| = n = n_u + n_w$. Then

$$\sigma(u; G) = \sigma(u; U) \cdot (w; W) \cdot \binom{n-1}{n_u - 1}. \qquad \square$$

Corollary 2.21   For any tree T, v is an accretion center node if and only if it is a centroid node.   □

The Betweenness Center

Suppose the nodes of a graph G are labeled $v_1, v_2, \ldots, v_n$. For each pair of nodes v_i and v_j, the *i-j betweenness value* $b_{ij}(v_k)$ of a node v_k is the ratio of the number of v_i-v_j geodesics which contain v_k to the total number of v_i-v_j geodesics. The *betweenness value* $C_B(v_k)$ of v_k is the sum of the numbers $b_{ij}(v_k)$ over all pairs i, j. The *betweenness center* $BC(G)$ of G is the set of all nodes with maximum betweenness value; a node in $BC(G)$ is a *betweenness center node*. The betweenness value of a node indicates its potential for control of communication among nodes in the graph. Hage and Harary [HH2] used betweenness centrality proposed by Freeman [F10] to analyze the structure of exchange networks among the Caroline Islands in the Pacific.

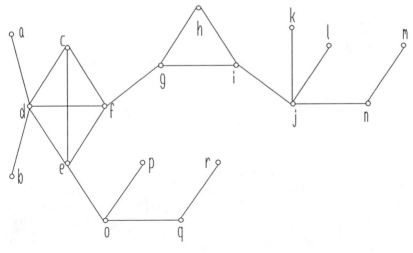

Figure 2.16   A graph to exercise generalized centers with.

EXERCISES 2.5

1. Find a formula for the cutting number of a node v in tree T in terms of the number of nodes in each branch at v.

2. Determine the cutting number of each node in the graph of Figure 2.16. What is the cutting center of G?

3. Find all 2-centers and all 3-centers of the graph in Figure 2.16.

4. Find all 2-medians of the graph in Figure 2.16.

5. Determine the 2-centricity of each node in Figure 2.16. What is the 2-centrum of G?

6. Find the accretion center of the graph of the graph in Figure 2.16. (Hint: Use Theorem 2.21 to help calculate $\sigma(v)$.)

7. Determine the betweenness value for each node in the graph of Figure 2.16. What is the betweenness center?

8. The number of sequential labelings beginning at v, $\sigma(v; G)$, is $(p-1)!$ if and only if deg $v = p - 1$.   (Slater [S17])

9. Let T be a tree and k be an integer, $0 \le k \le r(T)$. Then the k-ball branch weight centroid of T consists of a single node or a pair of adjacent nodes.   (Reid [R4])

Further Results

1. The maximum number of edges in a graph with radius r is

$$\begin{cases} p(p-1)/2 & \text{if } r = 1; \\ p(p-2)/2 & \text{if } r = 2; \text{ and} \\ (p^2 - 4rp + 5p + 4r^2 - 6r)/2 & \text{if } r \ge 3. \end{cases}$$

(Vizing [V1])

2. Let $a/b \le 1$ be a positive rational number in lowest terms. Determine the minimum number of nodes and edges in a graph G such that $|C(G)|/p = a/b$. (Note: There are several cases to consider: $b = a + 1$; $b \ne a + 1$ and $a = 1$ or 2;   or $b \ne a + 1$ and $a > 2$).

(Buckley [B23])

3. For each node v in a self-centered u.e.n. graph G with radius r, there are at least as many nodes at distance $r - 1$ from v as there are adjacent to v.   (Nandakumar and Parthasarathy [NP1])

4. The path centroid of a tree T contains the cutting center of T.

(Slater [S16])

5. The *path number of a node* v in a connected graph G is the number of paths that pass through v. For any tree T, the set of nodes with maximum path number is the cutting center of T. Also, for all $n \ge 3$,

there is a tree T having just one node with maximum path number n.   (Chinn [C9])

6. Let $N_k(v)$ be the set of nodes whose distance is at most k from v. Let $\rho_k(v) = \sum\{d(v, N_k(w)) : w \in V(G)\}$. The *k-nucleus* of a connected graph G to be the set of nodes which minimize $\rho_k(v)$. Then the k-nucleus of a tree T consists of either a single node or a pair of adjacent node if $0 \le k \le r(T)$.   (Slater [S18])

Intuitively, one feels that asymptotically half of all trees are central and half bicentral. Szekeres [S26] confirmed this for labeled trees; however, the problem remains open for unlabeled trees.

Buckley and Lewinter [BL3] obtained some relationships between the periphery and the set of eccentric nodes of a graph. For example, they determined tight bounds on the possible value of the diameter in terms of the radius for graphs in which no node is eccentric. Gu and Reid [GR1] extend that study to obtain all possible set inclusion relationships between the two sets. Earlier Bielak and Sysło [BS3] completed an analogous examination for the relationships between the center and periphery.

Self-centered graphs were surveyed recently in [B28]. Earlier Plesník [P5] completed a survey of results on the median of graphs and digraphs.

We have described embedding problems for centers and for medians. Holbert [H25] provides constructions combining the two problems. In particular, she shows that for graphs F and G and integer k, there is a connected graph H such that $\langle C(H) \rangle = F$, $\langle M(H) \rangle = G$, and the distance between F and G in H is k.

CHAPTER 3

Connectivity

Computer and telecommunication networks are often modeled by graphs. It is useful to know the *reliability* of a telecommunications network. That is, if one or two pieces of equipment fail, is it still possible for communication to proceed uninterupted? Network realiability problems are modeled by graphical networks where a number associated with each node and each edge represents the probability that the piece of hardware or connecting lines will fail. Networks and reliability will be studied in Chapter 12. A related concept is *vulnerability*, which is the susceptibility of a network to successful attack by adversaries. Both reliability and vulnerability are concerned with the subject of this chapter.

The connectivity of a graph is a particularly intuitive area of graph theory and extends the concepts of cutnode, bridge, and block. Two invariants called connectivity and edge-connectivity are useful in deciding which of two graphs is "more connected".

There is a rich body of theorems concerning connectivity. Many of these are variations of a classical result of Menger, which involves the number of disjoint paths connecting a given pair of nodes in a graph. We will see that several such variations have been discovered in areas of mathematics other than graph theory.

3.1   CONNECTIVITY AND EDGE-CONNECTIVITY

The *connectivity* $\kappa = \kappa(G)$ of a graph G is the minimum number of
nodes whose removal results in a disconnected or trivial graph. Thus the
connectivity of a disconnected graph is 0, while the connectivity of a
connected graph with a cutnode is 1. The complete graph K_p cannot be
disconnected by removing any number of nodes, but the trivial graph
results after removing $p - 1$ nodes; therefore, $\kappa(K_p) = p - 1$.

Analogously, the *edge-connectivity* $\kappa' = \kappa'(G)$ of a graph G is the
minimum number of edges whose removal results in a disconnected or
trivial graph. Thus $\kappa'(K_1) = 0$ and the edge-connectivity of a disconnected graph is 0, while that of a connected graph with a bridge is 1.
Connectivity, edge-connectivity and minimum degree are related by an
inequality due to Whitney [W3].

Theorem 3.1   For any graph G, $\kappa(G) \leq \kappa'(G) \leq \delta(G)$.

Proof   We first verify the second inequality. If G has no edges, then
$\kappa' = 0$. Otherwise, a disconnected graph results when all the edges incident with a node of minimum degree are removed. In either case $\kappa' \leq \delta$.

To obtain the first inequality, various cases are considered. If G is disconnected or trivial, then $\kappa = \kappa' = 0$. If G is connected and has a bridge e,
then $\kappa' = 1$. In this case $\kappa = 1$ since either G has a cutnode incident with
e or G is K_2. Finally, suppose G has $\kappa' \geq 2$ edges whose removal disconnects it. Clearly, the removal of $\kappa' - 1$ of these edges produces a bridge
$e = uv$. For each of these $\kappa' - 1$ edges, select an incident node different
from u or v. If the removal of these nodes produces a disconnected graph
then $\kappa < \kappa'$; if not, then $e = uv$ is a bridge, and hence the removal of u
or v will result in either a disconnected or trivial graph, so $\kappa \leq \kappa'$ in every
case.   □

Chartrand and Harary [CH3] constructed a family of graphs with
prescribed connectivities and minimum degree. This result shows that the
restrictions on κ, κ', and δ imposed by Theorem 3.1 cannot be improved.
For example, in Figure 3.1, we have $\kappa = 2$, $\kappa' = 3$, and $\delta = 4$.

Theorem 3.2   For all integers a, b, c such that $0 < a \leq b \leq c$, there
exists a graph G with $\kappa(G) = a$, $\kappa'(G) = b$, and $\delta(G) = c$.   □

As usual, $\lfloor x \rfloor$ denotes the greatest integer less than or equal to the
real number x, and $\lceil x \rceil$ is the least integer greater than or equal to x.

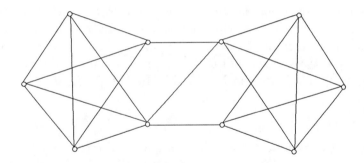

Figure 3.1   A graph with $0 \leq \kappa \leq \kappa' \leq \delta$.

Chartrand [C6] pointed out that if δ is large enough, then the second inequality in Theorem 3.1 becomes an equality.

Theorem 3.3   If a graph G has p nodes and the minimum degree $\delta(G) \geq \lfloor p/2 \rfloor$, then $\kappa'(G) = \delta(G)$.   □

For example, if G is regular of degree $r \geq \lfloor p/2 \rfloor$, then $\kappa'(G) = r$. In particular $\kappa'(K_p) = p-1$. Boesch and Suffel [BS5,6] extended Theorem 3.3 by finding a list of six necessary and sufficient conditions for the existence of a graph G with given values of p, κ, κ', and δ.

The analogue of Theorem 3.3 for connectivity does not hold. The problem of determining the largest connectivity possible for a graph with a given number of nodes and edges was proposed by Berge [B5] and the solution given by Harary [H12].

Theorem 3.4   Among all graphs with p nodes and q edges, the maximum connectivity is 0 when $q < p - 1$, and is $\lfloor 2q/p \rfloor$ when $q \geq p - 1$.

Proof   (Outline). Since the sum of the degrees of the nodes in any (p,q)-graph is $2q$, the average degree is $2q/p$. Therefore, $\delta(G) \leq \lfloor 2q/p \rfloor$, so $\kappa(G) \leq \lfloor 2q/p \rfloor$ by Theorem 3.1. To show that this value can actually be attained, an appropriate family of graphs can be constructed.   □

The same construction also gives those (p,q)-graphs with maximum edge-connectivity.

Corollary 3.4   The maximum edge-connectivity of a (p,q)-graph equals the maximum connectivity.   □

As noted by Bermond, Bond, Paoli, and Peyrat [BBPP1], the maximum connectivity and maximum edge-connectivity graphs are of fundamental importance in the design of minimum cost networks having uniform reliability. An excellent discussion of maximum connectivity graphs and network reliability can be found in Boesch [B11,12].

n-Connected Graphs

A graph G is *n-connected* if $\kappa(G) \geq n$ and *n-edge connected* if $\kappa'(G) \geq n$. Thus a nontrivial graph is 1-connected if and only if it is connected and 2-connected if and only if it is nonseparable and has more than one edge. So K_2 is the only nonseparable graph that is not 2-connected. From Theorem 1.6, it therefore follows that G is 2-connected if and only if every two nodes of G lie on a cycle. Dirac [D6] extended this observation to n-connected graphs.

Theorem 3.5   If G is n-connected, $n \geq 2$, then every set of n nodes of G lie on a cycle.   □

By taking G to be the cycle C_n, it is seen that the converse is not true for $n > 2$. However, a characterization of 3-connected graphs was given by Tutte [T15].

Theorem 3.6   A graph G is 3-connected if and only if it is a wheel or can be obtained from a wheel by a sequence of operations of the following two types:

1.   The addition of a new edge.

2.   The replacement of a node v having degree at least 4 by a pair of adjacent nodes v_1, v_2 such that in the resulting graph, each node is joined to exactly one of v_1 and v_2 and deg $v_1 \geq 3$ and deg $v_2 \geq 3$.   □

We discuss characterizations of n-connected and n-edge connected graphs in Section 3.2.

Connectivity of Line Graphs

Perhaps because the edge-connectivity of a graph G equals the connectivity of its line graph $L(G)$, line graphs have been studied extensively with respect to connectivity questions. The following observation of Chartrand and Stewart [CS2] was one of the first such results.

Theorem 3.7   If G is n-connected, $n \geq 2$, then its line graph $L(G)$ is also n-connected.

Proof   Let G be n-connected and suppose that $\kappa(L(G) < n$. Then removing $k < n$ nodes from $L(G)$ will produce either a disconnected or trivial graph. Each node of $L(G)$ corresponds with an edge of G with two nodes of $L(G)$ adjacent if and only if the corresponding edges of G are incident. Thus, by removing $k < n$ edges from G, we can produce either a disconnected or trivial graph, that is, $\kappa'(G) < n$. But Theorem 3.1 then implies $\kappa < \kappa' < n$, a contradiction. So $\kappa(L(G)) \geq n$, that is, $L(G)$ is n-connected. $\square$

Capobianco and Molluzzo [CM1] asked whether for each pair of integers a, b, with $1 < a < b$, there exists a graph G with $\kappa(G) = a$ and $\kappa(L(G)) = b$. Bauer and Tindell [BT1] constructed appropriate graphs to obtain the affirmative answer.

Theorem 3.8   For all integers a and b, $1 < a < b$, there is a graph G such that $\kappa(G) = a$ and $\kappa(L(G)) = b$. $\square$

EXERCISES 3.1

1. The connectivity of

    a. The octahedron $\overline{K}_2 + C_4$ is 4.

    b. The square of the cycle C_p, $p > 5$, is 4.

2. Every n-connected graph has at least $pn/2$ edges.

3. Construct a graph with $\kappa = 3$, $\kappa' = 4$, $\delta = 5$.

4. Given a positive integer n, construct a graph having $\kappa = n$, $\kappa' = n + 1$, $\delta = n + 2$.

5. Determine κ and κ' for each of the following:
    a. C_p b. K_p c. $W_{1,n}$ d. $K_{m,n}$.

6. If G is cubic, then $\kappa = \delta$.

7. Theorem 3.3 does not hold if $\kappa'(G)$ is replaced by $\kappa(G)$.

8. There is no 3-connected graph with seven edges.

9. If G is r-regular and $\kappa = 1$, then $\kappa' \leq \lfloor r/2 \rfloor$.

10. If G has diameter at most 2, then $\kappa'(G) = \delta(G)$.

11. If G is connected, then with the minimum taken over $v \in V(G)$, $\kappa(G) = 1 + \min \kappa(G - v)$.

12. If $\kappa'(G) \geq n$, $n \geq 2$, then $\kappa'(L(G)) \geq 2n - 2$.   (Zamfirescu [Z1])

13. If G is n-connected, then the join $G + K_1$ is $(n + 1)$-connected.

14. Use Tutte's Theorem 3.7 to show that the graph of the cube Q_3 is 3-connected.

3.2 MENGER'S THEOREM

In 1927 Menger [M5] showed that the connectivity of a graph is related to the number of disjoint paths joining two nodes. Many of the variations and extensions of Menger's which have since appeared have been graphical, and we discuss some of these here. Further variations of Menger's Theorem will be given in Section 3.3.

Let u and v be two nodes of a connected graph G. Two paths joining u and v are *disjoint* (sometimes called *node-disjoint*) if they have no nodes other than u and v (and hence no edges) in common; they are *edge-disjoint* if they have no edges in common. A set S of nodes, edges, or both *separates* u and v if u and v are in different components of $G - S$. Clearly, no set of nodes can separate two adjacent nodes. In Figure 3.2, we display a graph with two nonadjacent nodes s and t which can be separated by removing three nodes but no fewer. The classical theorem of Menger guarantees the existence of three node-disjoint paths joining s and t.

Theorem 3.9 The minimum number of nodes separating two nonadjacent nodes s and t equals the maximum number of disjoint s-t paths.

Proof We follow the elegant proof of Dirac [D8]. It is clear that if k nodes separate s and t, then there can be no more than k disjoint paths joining s and t.

It remains to show that if it takes k nodes to separate s and t in G, there are k disjoint s-t paths in G. This is certainly true if $k = 1$. Assume it is not true for some $k > 1$. Let h be the smallest such k, and let F be a graph with the minimum number of nodes for which the theorem fails for h. We remove edges from F until we obtain a graph G such that h nodes are required to separate s and t in G but for any edge e in G, only $h - 1$ nodes are required to separate s and t in $G - e$. We first investigate the properties of G.

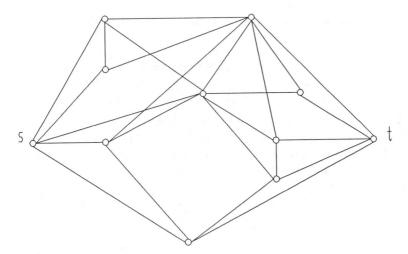

Figure 3.2   A graph illustrating Menger's Theorem.

By the definition of G, for any edge e of G there exists a set $S(e)$ of $h - 1$ nodes which separates s and t in $G - e$. Now $G - S(e)$ contains at least one s-t path, since it takes h nodes to separate s and t in G. Each such s-t path P must contain the edge $e = uv$ since P is not a path in $G - e$. So $u, v \notin S(e)$ and if $u \neq s, t$, then $S(e) \cup u$ separates s and t in G.

If there is a node w adjacent to both s and t in G, then $G - w$ requires $h - 1$ nodes to separate s and t and so it has $h - 1$ disjoint s-t paths. Replacing w, we have h disjoint s-t paths in G. Thus we have shown:

(1) No node is adjacent to both s and t in G.

Let W be any collection of h nodes separating s and t in G. An s-W path is a path joining s with some $w_i \in W$ and containing no other node of W. Call the collection of all s-W paths and $W - t$ paths P_s and P_t, respectively. Then each $s - t$ path begins with a member of P_s and ends with a member of P_t, because every such path contains a node of W. Moreover, the paths in P_s and P_t have only the nodes of W in common, since it is clear that each w_i is in at least one path in each collection and, if some other node were in both an s-W and a W-t path, then there would be an s-t path containing no node of W. Finally, either $P_s - W = \{s\}$ or $P_t - W = \{t\}$ since, if not, then both P_s plus the edges $\{w_1 t, w_2 t, \ldots\}$ and P_t plus the edges $\{s w_1, s w_2, \ldots\}$ are h-connected graphs with fewer nodes than G in which s and t are nonadjacent, and therefore in each there are h disjoint $s - t$ paths. Combining the s-W and W-t portions of these paths, we can construct h disjoint s-t paths in G, and thus have a contradiction. Therefore we have proved:

(2) Any collection W of h nodes separating s and t is adjacent either
to s or to t.

Now we can complete the proof. Let $P = \{s, u_1, u_2, \ldots, t\}$ be a shortest s-t path in G and let $u_1 u_2 = x$. Note that by (1), $u_2 \neq t$. Form
$S = \{v_1, v_2, \ldots, v_h - 1\}$ as above, separating s and t in $G - x$. By (1),
$u_1 t \notin G$, so by (2), with $W = S(x) \cup \{u_1\}$, $s v_i \in G$, for all i. Thus by (1),
$v_i t \notin G$, for all i. However, if we pick $W = S(x) \cup \{u_2\}$ instead, we have
by (2) that $s u_2 \in G$, contradicting our choice of P as a shortest s-t path,
and completing the proof of the theorem.   □

Chronologically, a first corollary of Menger's Theorem was published
by Whitney in a paper [W3] in which he included a criterion for a graph
to be n-connected.

Corollary 3.9   A graph G is n-connected if and only if every pair of
nodes are joined by at least n node-disjoint paths.   □

To indicate how Whitney's result follows from Theorems 3.9 we use
the concept of local connectivity. The *local connectivity* of two nonadjacent nodes u and v of a graph is denoted by $\kappa(u, v)$ and is defined as
the smallest number of nodes whose removal separates u and v. In these
terms, Menger's Theorem asserts that for any two specific nonadjacent
nodes u and v, $\kappa(u, v) = \mu_0(u, v)$, the maximum number of node-disjoint
paths joining u and v. Obviously, both theorems hold for complete graphs.
If we are dealing with a graph G which is not complete, then the observation which links Theorem 3.9 and its corollary is that $\kappa(G) = \min \kappa(u, v)$
over all pairs of nonadjacent nodes u and v.

Strangely enough, the theorem analogous to Theorem 3.9 in which
the pair of nodes are separated by a set of edges was not discovered until
much later. There are several nearly simultaneous discoveries of this result
which appeared in papers by Ford and Fulkerson [FF1] (as a special case
of their "max-flow, min-cut" theorem) and Elias, Feinstein, and Shannon
[EFS1].

Theorem 3.10   For any two nodes of a graph, the maximum number
of edge-disjoint paths joining them equals the minimum number of edges
which separate them.   □

Referring again to Figure 3.2, we see that s and t can be separated
by the removal of 5 edges but no fewer, and that the maximum number
of edge disjoint s-t paths is 5.

Table 3.1 Menger's Theorem Variations

Theorem	Objects separated	Maximum number	Minimum number
T. 3.9	specific u, v	disjoint paths	nodes separating u, v
C. 3.9	general u, v	disjoint paths	nodes separating u, v
T. 3.10	specific u, v	edge-disjoint paths	edges separating u, v

Even with only these three theorems available, we can see the beginnings of a scheme for classifying them. The difference between Theorems 3.9 and Corollary 3.9 may be expressed by saying that Theorem 3.9 involves two specific nodes of a graph, while Corollary 3.9 gives a bound in terms of two general nodes. This distinction as well as the obvious one between Theorems 3.9 and 3.10 is indicated in Table 3.1.

Thus we see that with no additional effort we can get another variation of Menger's Theorem by stating the edge form of Whitney's result.

Theorem 3.11   A graph is n-edge-connected if and only if every pair of nodes are joined by at least n edge-disjoint paths.   □

In Menger's original paper, there also appeared the following variation involving sets of nodes rather than individual nodes.

Theorem 3.12   For any two disjoint nonempty sets of nodes V_1 and V_2, the maximum number of disjoint paths joining V_1 and V_2 is equal to the minimum number of nodes which separate V_1 and V_2.   □

Of course it must be specified that no node of V_1 is adjacent with a node of V_2 for the same reason as in Theorem 3.9. Two paths joining V_1 and V_2 are understood to be disjoint if they have no nodes in common other than their endnodes. A proof of the equivalence of Theorems 3.9 and 3.12 is extremely staighforward and only involves shrinking the sets of nodes V_1 and V_2 to individual nodes.

EXERCISES 3.2

1. Construct a family of (p, q)-graphs with $2q/p$ integral such that $\kappa = 2q/p$.

2. State the result analogous to Theorem 3.9 for the maximum number of disjoint paths joining two adjacent nodes of a graph.

3. Every cubic triply-connected graph can be obtained from K_4 by the following construction. Replace two distinct edges u_1v_1 and u_2v_2 ($u_1 = u_2$ is permitted) by the subgraph with two new nodes w_1, w_2 and the new edges u_1w_1, w_1v_1, u_2w_2, w_2v_2, and w_1w_2.

4. Every block of a connected graph G is a wheel if and only if $q = 2p-2$ and $\kappa(u,v) = 1$ or 3 for any two nonadjacent nodes.

(Bollobás [B13])

5. Given two disjoint paths P_1 and P_2 joining two nodes u and v of a 3-connected graph G, is it always possible to find a third path joining u and v which is disjoint from both P_1 and P_2?

6. If G is n-connected, then the join $G + K_p$ is $(n + p)$-connected.

7. For the n-cube Q_n, $\kappa(Q_n) = \kappa'(Q_n) = n$ for all n.

8. For the cartesian product of two graphs G and H, we have $\kappa(G \times H) \geq \kappa(G) + \kappa(H)$.

3.3   PROPERTIES OF N-CONNECTED GRAPHS

A collection of paths of a graph G is an *independent set* if no two of them have a node in common. If each path in an independent set M of paths is an edge, then M is a *matching* for G. Thus a matching in a graph G is a set of edges of G, no two of which have a node in common. A number of variations of Menger's Theorem deal with matchings and independent paths. One such variation is the next theorem by Dirac [D7].

Theorem 3.13   A graph with at least $2n$ nodes is n-connected if and only if for any two disjoint sets V_1 and V_2 of n nodes each, there exist n independent paths joining these two sets of nodes.

Note that in this theorem these n independent paths do not have any nodes at all in common, not even their endnodes!

Proof   To show the sufficiency of the condition, we form the graph G' from G by adding two new nodes w_1 and w_2 with w_i adjacent to exactly the nodes of V_i, $i = 1, 2$. (See Figure 3.3.)

Since G is n-connected, so is G', and hence by Theorem 3.9 there are n disjoint paths joining w_1 and w_2. The restrictions of these paths to G are clearly the n independent V_1-V_2 paths we need.

To prove the other "half," let S be a set of at least $n - 1$ nodes which separates G into G_1 and G_2, with node sets V_1' and V_2', respectively. Then

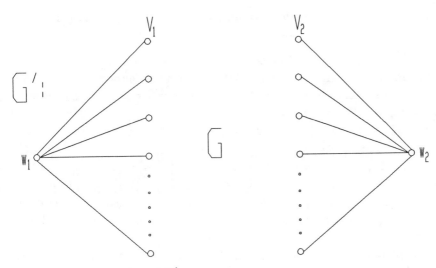

Figure 3.3　Construction of G'.

since $|V_1'| \geq 1, |V_2'| \geq 1$, and $|V_1'|+|V_2'|+|S| = |V| \geq 2n$, there is a partition of S into two disjoint subsets $|S_1|$ and $|S_2|$ such that $|V_1' \cup S_1| \geq n$ and $|V_2' \cup S_2| \geq n$. Picking any n-subsets V_1 of $V_1' \cup S_1$ and V_2 of $V_2' \cup S_2$, we have two disjoint sets of n nodes each. Every path joining V_1 and V_2 must contain a node of S, and since we know there are n independent V_1-V_2 paths, we see that $|S| \geq n$, and G is n-connected.　　　　□

Further Variations of Menger's Theorem

Fulkerson [F14] proved the following variation, which deals with cutsets rather than paths.

Theorem 3.14　In any graph, the maximum number of edge-disjoint cutsets of edges separating two nodes u and v is equal to the minimum number of edges in a path joining u and v, that is, the distance $d(u,v)$.□

We now describe several additional variations of Menger's theorem, all discovered independently and only later seen to be related to each other and to a graph theoretic formulation.

Let us define a *line of a matrix* as either a row or a column. Every entry of a *binary matrix* is 0 or 1. In a binary matrix M, a collection of lines is said to *cover* all the unit entries of M if every 1 is in one of these lines. Two 1's of M are called *independent* if they are neither in the same row nor in the same column. König [K3] obtained the next variation of Menger's Theorem in these terms.

Theorem 3.15 In any binary matrix, the maximum number of independent unit elements equals the minimum number of lines that cover all the units.   □

$$M = \begin{pmatrix} 0 & 0 & 1 & 0 & 0 & 0 \\ 1 & 1 & 0 & 1 & 0 & 1 \\ 0 & 0 & 1 & 0 & 0 & 1 \\ 0 & 1 & 1 & 0 & 1 & 0 \\ 0 & 0 & 1 & 0 & 0 & 1 \end{pmatrix} \qquad M' = \begin{pmatrix} 0 & 0 & 1 & 0 & 0 & 0 \\ 1 & 0 & 0 & 0 & 0 & 0 \\ 0 & 0 & 0 & 0 & 0 & 1 \\ 0 & 1 & 0 & 0 & 0 & 0 \\ 0 & 0 & 0 & 0 & 0 & 0 \end{pmatrix}$$

We illustrate Theorem 3.15 with the binary matrix M above. All the unit entries of M are covered by rows 2 and 4 and columns 3 and 6, but there is no collection of three lines of M which covers every 1. In the matrix M', four independent unit entries of M are shown and there is no such set of five units in M.

When this matrix M is regarded as an incidence matrix of sets versus elements, Theorem 3.15 becomes very closely related to the celebrated result of P. Hall [H3]. This provides a criterion for a family of finite sets $S_1, S_2, \ldots, S_m$ to possess a *system of distinct representatives*, i.e., a set $\{e_1, e_2, \ldots, e_m\}$ of distinct elements such that $e_i \in S_i$ for each i.

Theorem 3.16 There exists a system of distinct representatives for a family of sets $S_1, S_2, \ldots, S_m$ if and only if the union of any k of these sets contains at least k elements, for all k for 1 to m.   □

A node and an edge are said to *cover* each other if they are incident. A *node cover* of a graph G is a set of nodes which together covers all the edges of G. The next result due to König [K3] is equivalent to his Theorem 3.15.

Theorem 3.17 If G is bipartite, then the number of edges in a maximum matching equals the minimum number of nodes required to cover all the edges of G.   □

The problem of finding a maximum matching, the so-called matching problem, in a general graph will be discussed in detail in Chapter 11 on

Graph Algorithms. A matching that covers all the nodes of a graph G is called a *perfect matching* (also called a *1-factor*).

An additional Mengerian type theorem, the "max-flow", "min-cut" theorem due to Ford and Fulkerson will be discussed in detail in Chapter 12.

Node Degree and Connectivity

The following result of Bondy [B17] gives a condition on the degrees of the nodes which guarantees that a graph is n-connected.

Theorem 3.18   Let G be a graph of order $p \geq 2$ whose node degrees d_i satisfy $d_1 \leq d_2 \leq \cdots \leq d_p$. Let n be an integer, $1 \leq n \leq p - 1$. If

$$d_k \leq k + n - 2 \Longrightarrow d_{p-n+1} \geq p - k$$

for each k such that $1 \leq k \leq \lfloor (p - n + 1)/2 \rfloor$, then G is n-connected.

Proof   Suppose G satisfies the conditions of the theorem, but $\kappa(G) < n$. Then there exists a set S of at most $n-1$ nodes whose removal disconnects G. Consider the smallest component H of $G - S$ and call its order k. Then $k \leq \lfloor (p - n + 1)/2 \rfloor$ and the largest degree of a node in H is at most $k + n - 2 < p - k$. Thus $d_k \leq k + n - 2$ and the hypothesis of the theorem then implies that $d_{p-n+1} \geq p - k$. Since each node in $V(G) - V(H) - S$ has degree at most $p - k - 1$ and nodes in H also have degree less than $p - k$, only vertices in S have degree at least $p - k$. Now since $d_p \geq d_{p-1} \geq \cdots \geq d_{p-n+1} \geq p - k$, S contains at least n nodes, a contradiction. Thus, G is n-connected. $\square$

Boesch [B10] describes the result in Theorem 3.18 as the strongest monotone degree condition for n-connectedness of a graph. To see why, consider the graph G formed by joining each node of two disjoint complete graphs $H_1 \cong K_m$ and $H_2 \cong K_{p-m-n+1}$ to each node of $S \cong K_{n-1}$. For $u \in H_1$, $v \in H_2$, and $w \in S$, we have $\deg u = m+n-2$, $\deg v = p-m-1$, $\deg w = p - 1$ (see Figure 3.4). This graph "just fails" the conditions of Theorem 3.18 in the following sense. For each k such that $1 \leq k \leq \lfloor (p - (n - 1) + 1)/2 \rfloor$, we have

$$d_k \leq k + (n - 1) - 2 \Longrightarrow d_{p-(n-1)+1} \geq p - k,$$

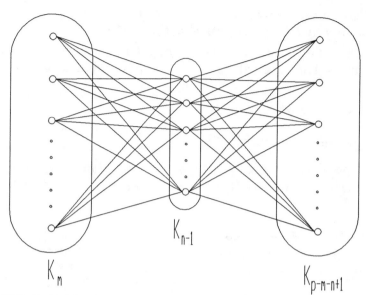

K_{n-1}

K_m

$K_{p-m-n+1}$

Figure 3.4 An "almost n-connected" graph.

so G is $(n-1)$-connected. However, when testing the condition of the theorem for n-connectivity with this graph, the hypothesis of the theorem holds for all k except for the single value $k = m$. Thus graph G is $(n-1)$-connected and "almost but not quite" n-connected.

EXERCISES 3.3

1. Let G be a graph of order $p \geq 2$, and let n be an integer where $1 \leq n \leq p-1$. If for every $v \in G$ we have $\deg v \geq \lfloor (p+n-2)/2 \rfloor$, then G is n-connected.

2. The minimum order of a node cover of G is at least $\delta(G)$.

3. Prove or disprove: Every node cover of G contains a minimum node cover.

4. If an n-connected graph G with p even is regular of degree n, then G has a perfect matching.   (Tutte [T13])

5. Prove the equivalence of Theorems 3.15 and 3.17.

6. If G is n-connected, $n \geq 2$, and $\delta(G) \geq (3n-1)/2$, then there exists a node $v \in G$ such that $G - v$ is n-connected.
                                        (Chartrand, Kaugers, and Lick [CKL2])

3.4   CIRCULANTS

In Section 3.1, we noted that among all graphs with p nodes and q edges, $q \geq p - 1$, the maximum connectivity is $\lfloor 2q/p \rfloor$ and this bound can always be attained. A chief reason for the importance of connectivity is its relation to the reliability and vulnerability of large-scale computer and telecommunication networks. (By vulnerability we mean susceptibility to successful attack by adversaries.) For example, on November 18, 1988 the major fiber-optic cable used for long distance calls on the east coast was inadvertently severed by a construction crew in New Jersey. This caused several hundred thousands of callers to receive a busy signal when the phone of the person being called actually was not in use. (See *New York Times*, Nov. 19, 1988, p. 1.)

Maximum connectivity graphs play an important role in the design of reliable networks. In this section we discuss a class of graphs known as circulants which contains those graphs. For a given positive integer, let $n_1, n_2, \ldots, n_k$ be a sequence of integers where
$$0 < n_1 < n_2 < \cdots < n_k < (p+1)/2.$$
Then the *circulant graph* $C_p(n_1, n_2, \ldots, n_k)$ is the graph on p nodes $v_1, v_2, \ldots, v_p$ with vertex v_i adjacent to each vertex $v_{i \pm n_j}$ (mod p). The values n_i are called *jump sizes*. For example, the circulant graphs $C_{10}(1, 3)$ and $C_{12}(1, 2, 5)$ are displayed in Figure 3.5.

Theorem 3.1 gives the relationship between $\kappa(G)$, $\kappa'(G)$, and $\delta(G)$ for an arbitrary graph G. Boesch and Tindell [BT2] characterized circulants for which $\kappa < \delta$.

Theorem 3.19   The circulants $C_p(n_1, n_2, \ldots, n_k)$ satisfies $\kappa < \delta$ if and only if for some proper divisor m of p, the number of distinct positive residues modulo m of the numbers $n_1, n_2, \ldots, n_k, p - n_k, \ldots, p - n_1$ is less than $\min\{m - 1, \delta m/p\}$. $\square$

Boesch and various coauthors considered both the size and structure of disconnecting sets of graphs (see [B12] or [BT3]). In one such study, Boesch and Wang [BW2] considered κ'-optimal graphs. Let $\kappa'^* = \max \kappa'(G)$ over all (p, q)-graphs g. Let Λ be the set of all (p, q)-graphs G for which $\kappa'(G) = \kappa'^*$. A graph $G^* \in \Lambda$ is κ'-*optimal* if it has the minimum number of disconnecting sets of edges of size κ' among all graph in Λ. Boesch and Wang showed that all circulants are κ'-optimal except the cycles and the graphs $C_{2n}(2, 4, \ldots, n - 1, n)$ for n odd.

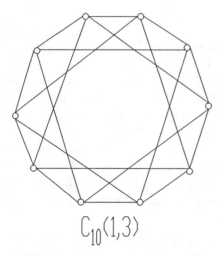

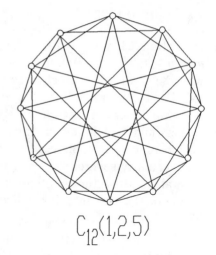

$$C_{10}(1,3)$$

$$C_{12}(1,2,5)$$

Figure 3.5 Two circulant graphs.

A regular graph with $\kappa = \delta$ for which the only minimum size disconnecting sets of nodes consists of the neighborhoods of single nodes is called a *super-κ graph*. Similarly, a regular graph with $\kappa' = \delta$ for which each minimum sized disconnecting sets of edges isolates a single node is called a *super-κ' graph*. Results for these graphs were described in the survey by Boesch and Tindell [BT3].

Diameter of Circulants

When designing a communication network, one not only wants to maximize the connectivity and edge-connectivity, but also to minimize the diameter as well as the number of edges. By minimizing the diameter, transmission times are kept small and the possibility of distortion due to a weak signal is avoided. Minimizing the number of edges will keep down the cost of building the network. Of course, one cannot have everything. That is, in general one cannot simultaneously maximize κ and κ' while minimizing $|E|$ and $d(G)$. To study this and similar problems, Harary [H15] introduced the concept of conditional invariants of a graph. These will be discussed in Chapter 6 where we explore extremal problems.

Determining the diameter of a circulant (or any graph for that matter) is a difficult problem. However, Boesch and Wang [BW3] were able to determine the minimum diameter among all circulants on p nodes and having two jump sizes.

Theorem 3.20   Let $G = C_p(m, m+1)$ be a circulant on p nodes with $p > 6$ and $m = \lceil(-1 + \sqrt{2p-1})/2\rceil$. Then $d(G) = m$ and m is the minimum diameter among all circulant graphs $C_p(n_1, n_2)$, where p is fixed, but n_1 and n_2 are arbitrary with $n_1 < n_2 < p/2$.   □

EXERCISES 3.4

1. A graph G is *minimal n-connected* if $\kappa(G) \geq n$ and for any edge $e \in G$, $\kappa(G - e) < n$. Every minimal n-connected graph has a node of degree at most $\lfloor 3n/2 \rfloor - 1$.
                            (Chartrand, Kaugers, and Lick [CKL2])

2. Every minimal 3-connected graph has two nodes of degree 3.
                            (Entringer and Slater [ES1])

3. Draw the circulants $C_{12}(1,3,4)$ and $C_{12}(2,4,5)$.

4. Use Theorem 3.19 to determine the minimum diameter of a circulant $C_{20}(n_1, n_2)$, where $n_1, n_2 < p/2$.

5. Determine all lists of jump sizes for which a circulant on 8 nodes is connected.

6. The complement of a circulant is a circulant.

7. The regular complete n-partite graph $K_{n,n,\ldots,n}$ is a circulant.

8. The circulants $C_7(1,2)$ and $C_7(1,3)$ are isomorphic.
                            (Turner [T11])

FURTHER RESULTS

1. Two elements of a lattice (see Birkhoff [B9]) are *incomparable* if neither dominates the other. By a *chain* in a lattice is meant a downward path from an upper element to a lower element in the "Hasse diagram" of the lattice. Prove: In any finite lattice, the maximum number of incomparable elements equals the minimum number of chains which include all the elements.   (Dilworth [D4])

A graph G is *k-critically n-connected* if for all $S \subset V(G)$ with $|S| \leq k$, we have $\kappa(G - S) = n - |S|$. This concept was introduced by Maurer and Slater [MS2]. A survey of results on these graphs was given by Mader [M1].

The *edge persistence* of a graph is the minimum number of edges that must be removed to increase its diameter. Exoo, Harary and Xu [EHX1] characterize graphs of diameter four with edge persistence equal to two.

Network reliability has been a very active area of research. So much so that a whole book (Colbourn [C16]) has been written on the subject.

CHAPTER 4

Hamiltonicity

One feature of graph theory that has helped popularize the subject lies in its applications to the area of puzzles and games. Sir William Hamilton suggested the class of graphs which bears his name with his invention of the Around the World Game in the mid 1800's. The game involved a solid dodecahedron, where each of the 20 nodes was labeled with the name of a well-known city. The object of the game was to find a round-the-world tour along the edges of the dodecahedron which passes through each city exactly once and returns to the initial city.

4.1 NECCESSARY OR SUFFICIENT CONDITIONS

In his Around the World Game, Hamilton asked for the construction of a cycle containing every node of a dodecahedron (its graph was given in Figure 1.7). If a graph G has a spanning cycle Z, then G is called a *hamiltonian graph* and Z a *hamiltonian cycle*. No elegant characterization of hamiltonian graphs exists, although several sufficient conditions are known as well as a few necessary conditions. Thus, there is no easy test to determine whether a given graph is hamiltonian. In fact, determining whether an arbitrary graph has a hamiltonian cycle is a standard example of an $\mathcal{NP}$-complete problem. More will be said about this in Chapter 11 on algorithms. The related *traveling salesman problem* will be discussed in Chapter 12.

The following theorem, due to Pósa [P8] gives a sufficient condition for a graph to be hamiltonian. It generalizes earlier results by Ore and Dirac which appear as its corollaries.

Theorem 4.1 If graph G has $p \geq 3$ nodes such that for every n, $1 \leq n < p/2$, the number of nodes of degree at most n is less than n, then G is hamiltonian.

Proof Assume the theorem does not hold and let G be a maximal non-hamiltonian graph with p nodes satisfying the hypothesis of the theorem. Obviously, the addition of any edge to a graph satisfying the hypotheses of the theorem results in a graph which also satisfies these conditions. Thus since the addition of any edge to G results in a hamiltonian graph by the maximality of G, any two nonadjacent nodes must be joined by a spanning path.

We first show that every node in G of degree at least $(p-1)/2$ is adjacent to every node of degree greater than $(p-1)/2$. Assume (without loss of generality) that $\deg v_1 \geq (p-1)/2$ and $\deg v_p \geq p/2$, but v_1 and v_p are not adjacent. Then there is a spanning path $v_1 v_2 \cdots v_p$ connecting v_1 and v_p. Let the nodes adjacent to v_1 be $v_{i_1}, v_{i_2}, \ldots, v_{i_n}$, where $n = \deg v_1$ and $2 = i_1 < i_2 < \cdots < i_n$. Clearly v_p cannot be adjacent to any node of G of the form $v_{i_j - 1}$ for otherwise there would be a hamiltonian cycle

$$v_1 v_2 \cdots v_{i_j - 1} v_p v_{p-1} \cdots v_{i_j} v_1.$$

Now since $n \geq (p-1)/2$, we have $p/2 \leq \deg v_p \leq p - 1 - n < p/2$ which is impossible, so v_1 and v_p must be adjacent.

It follows that if $\deg v \geq p/2$ for all nodes v, then G is hamiltonian. (This is stated below as Corollary 4.1b.) For the above argument implies that every pair of nodes in G are adjacent, so G is complete. But this is a contradiction since K_p is hamiltonian for all $p \geq 3$.

Therefore there is a node v in G with $\deg v < p/2$. Let m be the maximum degree among all such nodes and choose v_1 so that $\deg v_1 = m$. By hypothesis, the number of nodes of degree not exceeding m is at most $m < p/2$. Thus there must be more than m nodes having degree greater than m and hence at least $p/2$. Therefore there is some node, say v_p of degree at least $p/2$ not adjacent to v_1. Since v_1 and v_p are not adjacent, there is a spanning path $v_1 v_2 \cdots v_p$. As above, we write $v_{i_1}, v_{i_2}, \ldots, v_{i_m}$ as the nodes of G adjacent to v_1 and note that v_p cannot be adjacent to any of the m nodes v_{i_j-1} for $1 \leq j \leq m$. But since v_1 and v_p are not adjacent and v_p has degree at least $p/2$, m must be less than $(p-1)/2$, by the first part of the proof. Thus, by hypothesis, the number of nodes of degree at most m is less than m, and so at least one of the m nodes v_{i_j-1}, say v', must have degree at least $p/2$. We have thus exhibited two nonadjacent nodes v_p and v', each having degree at least $p/2$, a contradiction which completes the proof.   □

By specializing Pósa's Theorem, we obtain simpler but less powerful sufficient conditions due to Ore [O1] and Dirac [D5], respectively.

Corollary 4.1a   If $p \geq 3$ and for every pair u and v of nonadjacent nodes, $\deg u + \deg v \geq p$, then G is hamiltonian.   □

Corollary 4.1b   If for all nodes v of G, $\deg v \geq p/2$, then G is hamiltonian.   □

It is interesting to note that in the theory of hamiltonian graphs, stronger and stronger results have been obtained so that new results subsume older results as corollaries. Our next theorem, due to Chvátal [C13] can be used to prove Theorem 4.1.

Theorem 4.2   Let G be a graph of order $p \geq 3$ with degrees $d_1 \leq d_2 \leq \cdots \leq d_p$. If

(4.1) $d_i \leq i < p/2 \Longrightarrow d_{p-i} \geq p - i,$

then G is hamiltonian.

Proof   Assume the assertion is false and let G be a maximal nonhamiltonian graph which satisfies the hypothesis of the theorem. Let u and v be nonadjacent nodes for which $\deg u + \deg v$ is maximum with $\deg u \leq \deg v$. Since G is maximal, $G + uv$ is hamiltonian and there is a spanning u-v path $v_1 v_2 \cdots v_p$. As in Theorem 4.1, we write $v_{i_1}, v_{i_2}, \ldots, v_{i_m}$ as the nodes of G adjacent to v_1 and note that v_p cannot be adjacent to any of the m nodes v_{i_j-1} for $1 \leq j \leq m$ or we would find a hamiltonian cycle. This implies that $\deg v_1 + \deg v_p < p$, so $\deg v_1 < p/2$. Since for each j, $v_p v_{i_j-1} \notin E(G)$ and $\deg v_1 + \deg v_p$ is maximum among nonadjacent pairs of nodes, $\deg v_{i_j-1} \leq \deg v_1$. Let $\deg v_1 = m$. Then the number of nodes of degree at most m is at least m, so $d_m \leq m < p/2$. Thus (4.1) implies $d_{p-m} \geq p - m$. Thus each d_i, $p - m \leq i \leq p$, is at least $p - m$. Since $\deg v_1 = m$, v_1 is adjacent to at most m of these $m + 1$ nodes. Thus there is some node w not adjacent to $u = v_1$ with deg $w \geq p - m$. Thus deg u + deg $w \geq p >$ deg u + deg v, contradicting the choice of u and v. $\square$

It is interesting to compare this result and Bondy's Theorem 3.18 about connectivity. Given two degree sequences $d_1 \leq d_2 \leq \cdots \leq d_p$ and $d'_1 \leq d'_2 \leq \cdots \leq d'_p$, we say that the first sequence *majorizes* the second if $d_i \geq d'_i$ for each i. As with Bondy's Theorem, Theorem 4.2 is best possible in the following sense. If G is a graph with degree sequence $d_1 \leq d_2 \leq \cdots \leq d_p$ which does not satisfy (3.1), then there exists a nonhamiltonian graph G' whose degree sequence majorizes that of G.

Noting the similarity of Bondy's connectivity result and Chvátal's hamiltonicity result, it is not surprising that together they obtained several interesting results on hamiltonian graphs. The following theorem [BC1] is a sample of their collaboration.

Theorem 4.3 Let u and v be distinct nonadjacent nodes in a graph G of order p. If deg u + deg $v \geq p$, then $G + uv$ is hamiltonian if and only if G is hamiltonian. $\square$

One can then define the *closure* $cl(G)$ of a graph G as the graph obtained be recursively joining nonadjacent pairs of nodes whose degree sum is at least p. Then Theorem 4.3 implies that G is hamiltonian if and only if $cl(G)$ is hamiltonian.

Neighborhood Conditions

For a given node v, let $N(v)$ denote the set of neighbors of v. Then $|N(v)| =$ deg v. For a set $S \subset V(G)$, the neighborhood of S is the set $N(S) = \cup N(v)$ for $v \in S$. Recently, various authors have been studying hamiltonicity by considering the neighborhoods of nonadjacent nodes. The first observation they noted was that a minimum connectivity condition must be assumed to obtain any sort of reasonable results from the neighborhood sizes. Indeed, for a graph G on p nodes, one could have the size of the union of each pair of nonadjacent nodes as big as $p - 3$ and G still be disconnected, namely, with $G = K_{\lceil p/2 \rceil} \cup K_{\lfloor p/2 \rfloor}$. However, by simply assuming 2-connectedness, Faudree, Gould, Jacobson and Schelp [FGJS1] obtained a vast improvement on the existing degree conditions.

Theorem 4.4 If G is 2-connected and for every pair of nonadjacent nodes u and v, $|N(u) \cup N(v)| \geq (2p - 1)/3$, then G is hamiltonian. $\square$

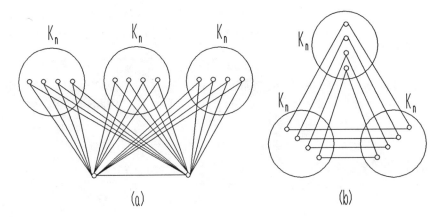

Figure 4.1   A nonhamiltonian and a hamiltonian graph.

To see that the bound given in Theorem 4.4 is sharp, consider the graph $G = 3K_n + K_2$ illustrated in Figure 4.1a. Graph G is 2-connected, has $p = 3n + 2$, and for all pairs of nonadjacent nodes u and v,

$$|N(u) \cup N(v)| = 2n = (2p - 1)/3 - 1,$$

but G is not hamiltonian. To see that Theorem 4.4 catches certain hamiltonian graphs that the previous sufficient conditions missed, consider the graph H in Figure 4.1b. Graph H is the cartesian product $K_n \times K_3$, which is perhaps easier to think of as 3 copies of K_n with a matching between coresponding pairs of nodes in each K_n. Graph H is hamiltonian but is not caught by any of Theorems 4.1-4.3 or their corollaries, since the degree of each node is only $p/3 + 1$. However, Theorem 4.4 implies that H is hamiltonian.

Theorem 4.4 has been improved upon in two different directions. In one case, Fraisse [F9] proved the following generalization conjectured in [FGJS1].

Theorem 4.5   If G is k-connected and for some $n \leq k$, every set S of n mutually nonadjacent nodes satisfies $|N(S)| > p(n - 1)/(p + 1)$, then G is hamiltonian.   □

So Theorem 4.4 is the special case of Theorem 4.5 with $k = n = 2$. The second improvement on Theorem 4.4 is due to Lindquester [L5]. She showed that Theorem 4.4 remains true when we restrict attention to nonadjacent nodes at distance two from one another.

Theorem 4.6 If G is 2-connected and every pair of nonadjacent nodes u and v with $d(u, v) = 2$ satisfies $|N(u) \cup N(v)| \geq (2p - 1)/3$, then G is hamiltonian.   □

Another result (which could be expressed in terms of neighborhoods but appears neater using degrees) is due to Fan [F1]. This result was in fact part of the inspiration for Theorem 4.6.

Theorem 4.7 If G is a 2-connected graph and every pair of nodes u and v with $d(u, v) = 2$ satisfies $\max\{\deg u, \deg v\} \geq p/2$, then G is hamiltonian.   □

Lesniak [L4] gives an interesting survey of neighborhood conditions, not only for hamiltonicity but for various other graphical properties as well.

EXERCISES 4.1

1. Give an example of a nonhamiltonian graph with 10 nodes such that for every pair of nonadjacent nodes u and v, $\deg u + \deg v \geq 9$.

2. How many spanning cycles are there in the complete bipartite graphs $K_{3,3}$ and $K_{4,3}$?

3. Pósa's theorem can be generalized as follows: Let G have order $p \geq 3$ and let $0 \leq k \leq p-2$. If for every integer i with $k+1 \leq i < (p+k)/2$, the number of nodes of degree not exceeding i is less than $i - k$, then every path of length k is contained in a hamiltonian cycle.

    (Kronk [K6])

4. If G is a (p, q)-graph with $p \geq 3$ and $q \geq (p^2 - 3p + 6)/2$, then G is hamiltonian. Furthermore, this bound is best possible.

    (Ore [O2])

5. If G is a bipartite graph with an odd number of nodes, then G is nonhamiltonian.

6. The bound $\deg v \geq p/2$ in Corollary 4.1b is best possible.

7. The n-cube Q_n is hamiltonian for $n \geq 2$.

8. The complete bipartite graph $K_{n,n}$ contains $n!(n-1)!/2$ hamiltonian cycles.

9. The closure of a graph is well defined, that is, $cl(G)$ is independent of the order in which one joins nonadjacent nodes whose degree sum exceeds p.

10. Give an example of a hamiltonian graph G with $\Delta(G) < p/2$ such that $cl(G) = G$.

11. If G has a pair of edge-disjoint hamiltonian cycles, then G has at least three hamiltonian cycles.   (Sloane [S21])

4.2   CONNECTIVITY AND HAMILTONICITY

Theorems 3.18 and 4.2 give some indication of a relationship between connectivity and hamiltonicity. In this section we describe several such links. A *theta graph* is a nonseparable graph with two nonadjacent nodes of degree 3 and all other nodes of degree two. Thus, a theta graph consists of two nodes of degree 3 and three disjoint paths joining them, with each path of length at least 2.

Theorem 4.8   Every hamiltonian graph is 2-connected. Every non-hamiltonian 2-connected graph has a theta subgraph.   □

It is easy to find a theta subgraph in the nonhamiltonian nonseparable graph of Figure 4.2.

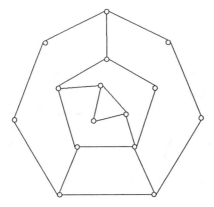

Figure 4.2   A nonhamiltonian nonseparable graph.

A set of nodes in G is *independent* if no two of them are adjacent. The largest number of nodes in such a set is the *node independence number*

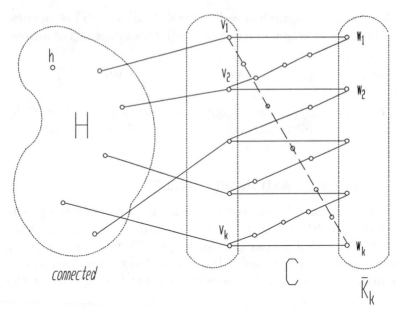

Figure 4.3   An illustration for Theorem 4.9.

of G and is denoted by $\beta(G)$ or β. The next result, due to Chvátal and Erdős [CE1] shows that if β is not too large then G is hamiltonian.

Theorem 4.9   Let G be a graph of order $p \geq 3$. If $\kappa(G) \geq \beta(G)$, then G is hamiltonian.

Proof   If $\beta(G) = 1$ then G is complete, so $p \geq 3$ implies that G is hamiltonian. Thus assume that $\beta(G) \geq 2$. Since $\kappa(G) \geq \beta(G) \geq 2$, G contains at least one cycle. Let C be a longest cycle in G and suppose that C is not a spanning cycle. Let H be a connected component of $G - C$ and let $v_1, v_2, \ldots, v_k$ be nodes of C which are adjacent to nodes of H. No two of the nodes v_i, v_j are adjacent in C, since otherwise we could replace the edge $v_i v_j$ of C by a path through H and obtain a longer cycle than C. Now traverse C in some fixed direction and let w_i be the node following v_i in the traversal. Then no node w_i is adjacent to a node of H for otherwise a replacement of edge $v_i w_i$ by a path through H would produce a longer cycle. Also, the nodes w_i are independent since if two of them, say w_i and w_j, are adjacent, then a longer cycle than C exists. This cycle consists of the edges of C with $v_i w_i$ and $v_j w_j$ deleted, plus the edge $w_i w_j$, and a $v_i - v_j$ path through H (see Figure 4.3).

Let h be a node of component H. Then $\{h, w_1, w_2, \ldots, w_k\}$ is an independent set. Since $\{v_1, v_2, \ldots, v_k\}$ contains all the nodes of C which are adjacent to nodes of H, $G - \{v_1, v_2, \ldots, v_k\}$ is disconnected. Thus $\kappa(G) \leq k$. But since $\{h, w_1, w_2, \ldots, w_k\}$ is an independent set, $\beta(G) \geq k + 1$. Therefore, $\beta(G) > \kappa(G)$, a contradiction. Thus, C is spanning cycle and G is hamiltonian. $\Box$

The condition in Theorem 4.9 is certainly not necessary. For example, the cycle C_p has $\kappa = 2$ and $\beta = \lfloor p/2 \rfloor$ but is hamiltonian. Bondy [B19] showed that if G has order $p \geq 3$ and $\deg u + \deg v \geq p$ for all pairs of distinct nonadjacent nodes, then $\kappa(G) \geq \beta(G)$. Thus, Theorem 4.9 implies Ore's Theorem. It can also be used to prove the following result of Nash-Williams [N2].

Theorem 4.10   Every k-regular graph $(k \geq 2)$ of order $2k + 1$ is hamiltonian. $\Box$

EXERCISES 4.2

1. If every two nodes of G are joined by a spanning path, then G is 3-connected.

2. Prove or disprove: If a graph G contains an induced theta subgraph, then G is not hamiltonian. If false, find the smallest counterexample.

3. If G is a graph with $p \geq 3$ nodes such that the removal of any set of at most n nodes results in a nontrivial hamiltonian graph, then G is $(n + 2)$-connected.   (Chartrand, Kapoor, Lick [CKL1])

4. If G is bipartite with each part of order n and if its size q satisfies $q \geq n^2 - n + 2$, then G is hamiltonian.

5. If G has order $p \geq 3$ and contains n nodes of degree $p - 1$, then $n \geq \beta_0(G)$ implies that G is hamiltonian.

6. Every cubic hamiltonian graph has at least three spanning cycles.
                                                                    (Tutte [T12])

7. If G is hamiltonian and S is a set of k nodes of G, then $G - S$ has at most k components.

8. Draw a 3-connected planar nonhamiltonian graph on 11 nodes.

9. Do there exist nonhamiltonian graphs with arbitrarily high connectivity?

4.3   GRAPH OPERATIONS AND HAMILTONICITY

In Chapter 1 we defined a number of unary operations on a graph, such as the square, cube, complement, and line-graph of a given graph. These operations have served to obtain hamiltonicity results, some of which are given in this section.

Powers of graphs

Recall that the nth power G^n of G has the same node set as G and nodes u and v are adjacent in G^n if their distance in G is at most n.

If the diameter of G is k, then G^k is complete and therefore hamiltonian. However, it may be that G^n is hamiltonian for some $n < k$. Plummer and Nash-Williams independently conjectured that if G is 2-connected, then G^2 is hamiltonian. Their true conjecture was proved by Fleischner [F4] in a tour de force of mathematical reasoning. Říha [R5] recently obtained a simple proof of this result.

Theorem 4.11   If G is a 2-connected graph, then G^2 is hamiltonian.   □

Fleischner's Theorem can be used to obtain a number of results on generalizations of hamiltonicity which we discuss in the next section. For now we present the following simple result whose proof was kindly provided by C. Thomassen.

Theorem 4.12   If G has $p \geq 3$, then G^2 or $(\overline{G})^2$ is hamiltonian.

Proof   If G is 2-connected, then Theorem 4.11 implies that G^2 is hamiltonian. If G is disconnected, then $(\overline{G})^2 = K_p$. If G and $(\overline{G})$ are connected, and v is a cutnode in G, then $(\overline{G})^2 - v = K_{p-1}$. Since $\deg v$ in $(\overline{G})$ is at least 1, $\deg v$ in $(\overline{G})^2$ is at least 2, and hence $(\overline{G})^2$ is hamiltonian.   □

Corollary 4.12   If G is self-complementary, then G^2 is hamiltonian. □

Although the square of a connected graph of order at least 3 is not always hamiltonian, its cube is. The next result is due to Sekanina [S8].

Theorem 4.13   If G is a connected graph of order at least 3, then G^3 is hamiltonian.

Proof   Let G be a connected graph of order at least 3. We proceed by induction on the radius of G. If $r(G) = 1$, then $d(G) \leq 2$ so clearly G^3 is hamiltonian. Thus assume that for all graphs H with $r(H) < k$, H is hamiltonian. Clearly, if $d(G) \leq 3$, then G^3 is hamiltonian. Thus suppose that $r(G) = k \geq 2$, $d(G) \geq 4$, and let v be a central node of G. Then we may perform a breadth first search with root v to form a spanning tree T of G so that $r(T) = r(G)$.

Let P be the set of peripheral nodes of T. Each node $u \in P$ has degree 1 in T, so has a unique neighbor $n_T(u)$ in T. Let $T' = T - P$, and let $G' = \langle T' \rangle$. Then $r(G') \leq r(T') = r(T) - 1 = k - 1$, so by the inductive hypothesis, $(G')^3$ has a hamiltonian cycle C. We can expand C to form a hamiltonian cycle of G as follows. Let the nodes of C be labeled so that $C = v_1 v_2 \cdots v_t v_1$. For each i, $1 \leq i \leq t$, let P_i be the set of those nodes of P whose unique neighbor in T is v_i. For a given i, the nodes of P_i are at distance 2 from one another and from $v_{(i+1) \bmod t}$ in T so are adjacent in G^3. For each i, if $P_i \neq \emptyset$, delete edge $i, i+1 \pmod t$ from C include an edge from v_i to one node of P_i, followed by a path spanning P_i which ends in node p_i^*, followed by edge p_i^*, v_{i+1}. This procedure results in a cycle which spans G^3, which is therefore hamiltonian. $\square$

Eulerian Graphs

A graph G is *eulerian* (pronounced "oileerian") if it has a closed spanning trail which includes each edge. These graphs are named after Leonhard Euler (1707-1782) who became the father of graph theory as well as topology when in 1736 he settled a famous unsolved problem of his day called the Königsberg Bridge Problem. There were two islands linked to each other and to the banks of the Pregel River as shown in Figure 4.4a. The problem was to begin at any of the four land areas and walk across each bridge exactly once. Euler replaced each land area by a node and each bridge by an edge joining the corresponding nodes. This produces a *multigraph* (a graph allowing multiple edges, see Figure 4.4b).

Rather than treating this specific problem, Euler showed that no such traversal of the Königsberg bridges was possible when he proved the following more general result, often called the first theorem of graph theory.

Theorem 4.14   A connected graph G is eulerian if and only if each node of G has even degree. $\square$

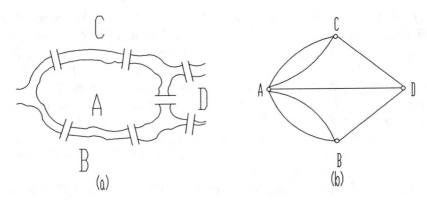

Figure 4.4 The Königsberg Bridges and corresponding multigraph.

Line Graphs

Recall that the line graph $L(G)$ of G has $V(L(G)) = E(G)$ and two nodes of $L(G)$ are adjacent if the corresponding edges of G are adjacent, i.e., have just one common node. Obviously, if G is eulerian, then $L(G)$ is hamiltonian. The iterated line graphs are $L^2(G) = L(L(G))$, $L^3(G) = L(L^2(G))$, etc. An early result on hamiltonian iterated line graphs is the following theorem of Chartrand and Wall [CW1].

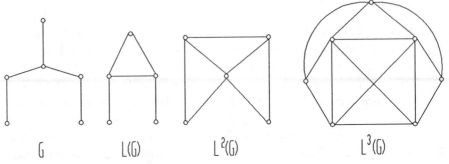

Figure 4.5 A sequence of iterated line graphs.

Theorem 4.15 If G is connected and $\delta(G) \geq 3$, then $L^2(G)$ is hamiltonian.

Proof If G is connected and $\delta(G) \geq 3$, then each pair of incident edges of G are mutually incident with a third edge. This means that in $L(G)$ every edge is contained in a triangle. Thus $L(G)$ has no bridges, each node of $L(G)$ has degree at least 4, and $L(G)$ has a spanning closed trail that is incident with each edge of $L(G)$. Let this trail T^* be $e_1, e_2, \ldots, e_n, e_1$. We can expand this to a list of edges which produces a hamiltonian path in $L(L(G))$ as follows. Begin with $e_1 = v_1 v_2$ and follow it by a list of all other edges not on T^* which are incident with v_2. The list can be in any order and should be followed by $e_2 = v_2 v_3$. In general, after listing $e_i = v_i v_{i+1}$, list all edges not on T^* which are incident with v_{i+1} and have not already been listed. End the list with the edge $e_n = v_n v_1$. Since each edge in the resulting list is incident with the edge preceding and following it, the first and last edges are incident, and the list contains every edge of $L(G)$, the corresponding path in $L(L(G))$ is a spanning path, so $L^2(G)$ is hamiltonian.   □

From Theorem 4.15, we obtain the following result due to Chartrand [C7].

Corollary 4.15   If G is connected and not a path, then $L^n(G)$ is hamiltonian for some n.   □

Thomassen [T9] conjectured that if $L(G)$ is 4-connected, then $L(G)$ is hamiltonian.

EXERCISES 4.3

1. Find a connected graph G whose square is not hamiltonian.

2. Construct a class of graphs to show that for an arbitrarily large positive integer n, there exist connected graphs G which are not paths such that $L^n(G)$ is not hamiltonian.

3. A graph G is eulerian if and only if G is connected and every block of G is eulerian.

4. A nontrivial connected graph G is eulerian if and only if every edge of G lies on an odd number of cycles.

5. Graph G is *randomly eulerian from node v* if every maximal trail starting at v produces an eulerian circuit. Prove: G is randomly eulerian from v if and only if it is eulerian and v is on every cycle of G.

6. A cycle C is a *dominating cycle* of G if each node of G is either in C or adjacent to a node in C. Let G be a graph without isolated nodes. Then $L(G)$ is hamiltonian if and only if G contains a dominating cycle or G is a star $K_{1,n}$ with $n \geq 3$.

(Harary and Nash-Williams [HN1])

7. If G is hamiltonian, then $L(G)$ is hamiltonian.

8. If G is eulerian, then $L(G)$ is both eulerian and hamiltonian.

9. If G is n-edge-connected, then

    a. $L(G)$ is n-connected, and

    b. $L(G)$ is $(2n - 2)$-edge-connected.

                                     (Chartrand and Stewart [CS2])

10. Prove or disprove: If G and H are hamiltonian, then their product $G \times H$ is hamiltonian.

11. Construct a connected graph G with $p \geq 4$ such that $L(G)$ is not eulerian but $L^2(G)$ is.

12. There is no connected graph G with $p \geq 5$ such that $L^2(G)$ is not eulerian and $L^3(G)$ is.

13. The line graph $L(G)$ is eulerian if and only if G is connected and the degrees of all nodes of G are of the same parity.

4.4   GENERALIZATIONS OF HAMILTONICITY

There has been no computationally useful characterization found for hamiltonian graphs, and it is the consensus among graph theorists that no such criterion can ever be found. However, related problems have been considered which have provided interesting questions for research. We now discuss several of these questions.

Highly Hamiltonian Graphs

If we know that a graph G is hamiltonian, it is still not an easy task to find a spanning cycle in G. As we are tracing the cycle, it does not matter where we start, but it almost always matters to which node we proceed next. However, for certain hamiltonian graphs it is easy to find a spanning cycle. A graph G is *randomly hamiltonian* if a hamiltonian cycle always results upon starting at any node of G and then successively proceeding to any adjacent node not yet chosen until no new nodes are available. These graphs were characterized by Chartrand and Kronk [CK1] as follows.

Theorem 4.16   A graph G with $p \geq 3$ nodes is randomly hamiltonian if and only if it is one of the graphs C_p, K_p, or $K_{n,n}$ with $p = 2n$. $\square$

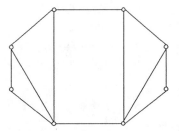

Figure 4.6   A pancyclic graph which is not node-pancyclic.

Another class of highly hamiltonian graphs are graphs G for which there is not only a cycle of length p in G, but a cycle of each smaller length as well. A graph G of order p is *pancyclic* if it contains a cycle of length k for each integer, $3 \leq k \leq p$. Bondy [B18] gave the following sufficient condition for a hamiltonian graph to be pancyclic.

Theorem 4.17   If G is a hamiltonian graph of order p and size $q \geq p^2/4$, then either G is pancyclic or p is even and G is $K_{p/2,p/2}$.   □

One could require even more than a graph having a cycle of each possible length. A hamiltonian graph G is *node-pancyclic* if for each node v, G contains a cycle of each length k, $3 \leq k \leq p$, which contains node v. Note that in a node-pancyclic graph, every node is contained in a triangle. The graph $(C_7)^2$ is an example of a node-pancyclic graph (note that it is also the circulant $C(1,2)$). In Figure 4.6, we display a graph of order 8 which is pancyclic but not node-pancyclic.

Nearly Hamiltonian Graphs

There are several classes of graphs that have been studied which are "almost" hamiltonian. The first of these are *traceable* graphs which have a spanning path but no spanning cycle. A special class of these graph has received considerable attention. A graph G is *hamiltonian-connected* if for each pair of nodes u, v, there is a spanning path joining u and v. It is easy to see that any hamiltonian-connected graph G of order at least 3 is hamiltonian. Simply consider two nodes u and v of G and let P be a spanning path joining u and v. Let w be the neighbor of u on P. Then G has a spanning path P' joining u and w, and $P' \cup uw$ is a spanning cycle of G.

It is interesting (but not surprising) that for each sufficient condition of Section 4.3 which guarantees that a graph G is hamiltonian, there is an analogous condition which shows that G is hamiltonian-connected. Since hamiltonian-connectedness is the stronger of the two, tighter restrictions are required to guarantee hamiltonian-connectedness. For example, analogous to Corollary 4.1a, we have the following result of Ore [O4].

Theorem 4.18   If G is a connected graph of order p such that for all pair of nonadjacent nodes u,v,

$$\deg u + \deg v \geq p + 1,$$

then G is hamiltonian-connected. $\square$

A concept that is similar in flavor to pancyclic graphs is that of pan-connected graphs. A hamiltonian-connected graph G is *panconnected* if for each pair of nodes u,v and each integer k, $d(u,v) \leq k < p$, there is a path of length k joining u and v. Williamson [W4] gave the following sufficient condition for a graph to be panconnected.

Theorem 4.19   If G has order $p \geq 4$, and if for each node v, $\deg v \geq p/2 + 1$, then G is panconnected.

Proof   If $p = 4$ or $p = 5$, then $G = K_4$ or K_5, respectively. Thus, assume the theorem is true for $p = t$, and show it true for $p = t + 2$. Let G be a graph on $p = t + 2$ nodes such that $\delta(G) \geq p/2 + 1$. Let u and v be any two nodes of G. Then u and v are each adjacent to at least half the nodes of G. The graph $G' = G - \{u, v\}$ has t nodes and $\delta \geq t$, so Corollary 4.1b implies that G' has a hamiltonian cycle $C' = u_1, u_2, \ldots, u_t$. The degree conditions on u and v in G then imply that $d(u,v) \leq 2$. Suppose that for some m, $2 \leq m \leq t$, node v does not have a neighbor whose distance on C' is $m - 2$ from a neighbor of u on C'. Then $\deg v < p/2 + 1$, a contradiction. Thus, the existence of such a neighbor for each given value of m guarantees a path of each length between $d(u,v)$ and $t + 2 = p$. Hence, G is panconnected. $\square$

It should be noted that Lesniak's survey paper [L4] also describes a number of neighborhood results relating to generalizations of hamiltonicity. Various researchers have determined sufficient conditions for a non-hamiltonian graph to contain a path or a cycle of some specified length. We shall discuss those results in Chapter 5.

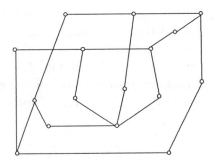

Figure 4.7 A nonhamiltonian graph.

EXERCISES 4.4

1. A graph is *randomly traceable* if a hamiltonian path always results upon starting at any node and successively proceding to an adjacent node not yet chosen until no new nodes are available. A graph G with $p \geq 3$ nodes is randomly traceable if and only if it is randomly hamiltonian.   (Chartrand and Kronk [CK1])

2. The smallest nonseparable graph whose line graph is not hamiltonian is the theta graph with 8 nodes in which the distance between the nodes of degree 3 is 3.

3. If G is a hamiltonian-connected graph of order at least 4, then $\delta(G) \geq 3$.

4. The Petersen graph is hamiltonian-connected, but not hamiltonian.

5. The graph in Figure 4.7 does not have a hamiltonian path.

6. There exists a pancyclic graph that is not panconnected.

7. The minimum number of edges in a hamiltonian-connected graph of order $p \geq 4$ is $\lfloor (3n + 1)/2 \rfloor$.   (Moon [M12])

8. If G is connected with order at least 4, then G^3 is panconnected.
                                           (Alavi and Williamson[AW1])

9. If G is hypohamiltonian and u and v are adjacent nodes on a cycle which spans $G - x$, then u and v are not both adjacent to node x.

FURTHER RESULTS

Jackson [J1] showed that every 2-connected, k-regular graph of order $p \leq 3k$ is hamiltonian. Liu, Yu, and Zhu [LYZ1] improved this bound to $k \geq p/3 - 1$ with the Petersen graph as one exception. Bondy and Kouider [BK1] give a simple proof of this result.

Several results relating hamiltonicity and connectivity concern planar graphs. Tait [T1] conjectured that every cubic 3-connected planar graph contains a spanning cycle. Tutte [T12] disproved this by showing that the 3-connected planar graph with 46 nodes in Fig. 4.8 is not hamiltonian. Tutte [T14] also showed that every 4-connected planar graph is hamiltonian. Thomassen [T8] showed that it is even hamiltonian-connected.

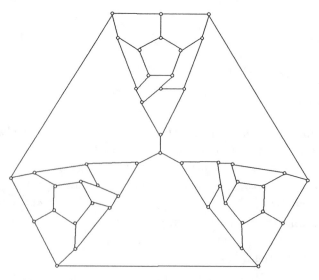

Figure 4.8   The Tutte graph.

A graph G is *hamiltonian-connected from node v* if there is a v-u spanning path for each $u \in G - v$. Chartrand and Nordhaus [CN1] showed that such graphs have at least $\lceil (5p-1)/4 \rceil$ edges. Recently, Knickerbocker, Lock, and Sheard [KLS1] proved that bound to be sharp. Graph G is *uniquely hamiltonian-connected from v* if it is hamiltonian-connected from v and each v-u is unique. In a separate paper [KLS2], the same authors extend a result of Hendry [H22] to show that a graph of order $p > 3$ can be uniquely hamiltonian-connected from at most one node.

Another class of nearly-hamiltonian graphs are the hypohamiltonian graphs introduced by Gaudin, Herz, and Rossi [GHR1]. A nonhamiltonian graph G is *hypohamiltonian* if for every node v in G, the graph $G - v$ is hamiltonian. The Petersen graph is the smallest hypohamiltonian graph. Note that hypohamiltonian graphs exist for almost every possible order (see Chvátal [C14] and Thomassen [T4]). There are even planar, cubic 3-connected hypohamiltonian graphs [T7]. The smallest known such graph is displayed on the cover of Beineke and Wilson [BW1].

CHAPTER 5

Extremal Distance Problems

Let f be a real-valued function whose domain is the set of all graphs and let $\mathcal{P}$ be any graphical property. One could ask the question: "What is the largest (or smallest) possible value of f among all graphs with property P?" This is one paradigm of questions in extremal graph theory. As with topics discussed in previous chapters, extremal graph theory is an area on which a whole book could be written (and has been, see Bollobás [B15]). In this chapter we shall focus our attention on extremal problems relating to radius, diameter, and long paths and cycles in graphs.

5.1 RADIUS

In Chapter 2 we defined the radius $r(G)$ of a connected graph G as the minimum eccentricity of its nodes and the center $C(G)$ as the set of all nodes v with $e(v) = r(G)$. We saw that the center plays an essential role in a number of facility location problems. Since $r(G)$ is a real valued function, in fact positive integer valued, the radius could play the role of the function f in the extremal graph paradigm. However, in many extremal problems, $r(G)$ instead plays a role in the property P. The following result of Vizing [V1] is an example where $r(G)$ plays this latter role.

Theorem 5.1   The maximum number of edges in a graph on p nodes with radius r is

$$
\begin{cases}
p(p-1)/2, & r = 1; \\
\lfloor p(p-2)/2 \rfloor, & r = 2; \\
(p^2 - 4pr + 5p + 4r^2 - 6r)/2 & r \geq 3.
\end{cases}
\qquad \square
$$

Vizing gives an interesting proof of this result which uses double induction. Extremal graphs for the three possibilities are K_p for $r = 1$, and $K_p - \{$a 1-factor$\}$ when p is even and $r = 2$. When $r = 2$ and p is odd, an extremal graph can be formed as follows. Begin with complete graphs $G = K_{(p-1)/2}$ and $H = K_{(p+1)/2}$. Join each node of H with $(p-3)/2$ nodes of G in such a way that no node of G achieves degree $p - 1$. An extremal graph for $r \geq 3$ is shown in Figure 5.1. This graph consists of a complete graph K_{p-2r} all of whose nodes are joined to three consecutive nodes on a cycle C_{2r}.

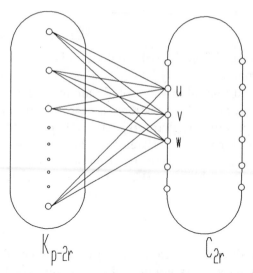

Figure 5.1   An extremal graph on p nodes with radius r.

Radius-Minimal and Radius-Critical Graphs

For any connected graph G, it is easy to generate a spanning tree T of G for which the distances from a fixed node v are preserved. One simply uses the well-known "breadth-first search" algorithm with root v. This algorithm begins at a node v and branches out to its neighbors u, including the edges uv in the tree. Next, edges joining those nodes at distance one from v with nodes at distance two from v are included so as not to form any cycles. This process continues until a spanning tree is formed. This process is illustrated in Figure 5.2 where the central node d is used as the initial node. For now it is important to note that if we begin at a central node, the spanning tree T will have the same radius as G. Such a tree is called a *radius-preserving spanning tree*. We shall discuss breadth-first search and other algorithms in detail in Chapters 11 and 12.

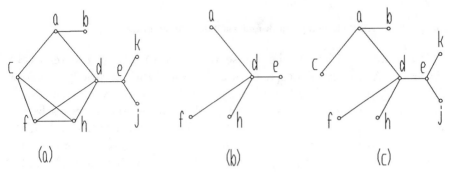

Figure 5.2   A breadth-first search at a node.

Vizing's result described in Theorem 5.1 concerns radius-maximal graphs. That is, if a graph G on p nodes has radius r and its number of edges equals the given bound, then any new edge added to G will necessarily decrease the radius. We now look at graphs at the opposite extreme. If one removes an edge from a graph G, it is clear that the radius may increase or stay the same, but it certainly could not decrease. A graph G is called *radius-minimal* if $r(G - e) > r(G)$ for every edge e in G. Gliviak [G7] characterized such graphs as follows.

Theorem 5.2   A nontrivial graph G is radius-minimal if and only if G is a tree.

Proof   If G is a tree, then clearly G is radius-minimal since the removal of any edge will disconnect G, resulting in an infinite radius.

If G is radius-minimal, then $r(G)$ must be finite so G is connected. Assume that G is connected and not a tree. Then G has a radius-preserving spanning tree T which necessarily has fewer edges than G. Thus it is possible to remove an edge from G without decreasing the radius. Hence a radius-minimal graph must be a tree.   □

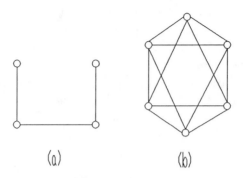

Figure 5.3 An r-changing graph and an r-decreasing graph.

Next we consider graphs whose radius is altered by the removal of any node. A nontrivial graph G is called r-critical, or briefly r-critical, if for every node v in G, $r(G - v) \neq r(G)$. Every even path P_{2n} is r-critical. By removing an endnode of P_{2n}, the radius decreases by one, but removing an internal node of P_{2n} makes the radius unbounded. It is a simple observation that if G is an r-critical graph and v is one of its nodes, then $r(G - v) < r(G)$ if and only if v is a peripheral node, and in this case $r(G - v) = r(G) - 1$. Gliviak [G7–8] obtained most of the results on these graphs.

The class of r-critical graphs can be partitioned into three sets:

   r-decreasing graphs for which $r(G - v) = r(G) - 1$ for all v;

   r-increasing graphs for which $r(G - v) > r(G)$ for all v; and

   r-changing graphs which comprise all other r-critical graphs.

Thus each r-changing graph contains at least one node v for which $r(G - v) > r(G)$ and one node u for which $r(G - u) < r(G)$. Figure 5.3(a) gives an example of an r-changing graph, while Figure 5.3(b) is an r-decreasing graph. Gliviak [G8] reduced the study of r-critical graphs to that of r-decreasing graphs by means of the following theorem.

Theorem 5.3   Every connected r-critical graph G is either r-decreasing or consists of an r-decreasing subgraph H, and endpaths so that one endpath of length $r(G) - r(H)$ is joined to each node of H.   □

Corollary 5.3   There are no r-increasing graphs.   □

In another paper, Gliviak [G7] characterized r-critical graphs having radius 2, as well as all r-decreasing graphs.

Theorem 5.4 A graph G of radius 2 is r-critical if and only if it is either the path P_4 or the complete multipartite graph $K(2,2,\ldots,2)$ with $n \geq 2$ parts. □

Note that P_4 is r-changing, while the complete multipartite graphs $K(2,2,\ldots,2)$ are r-decreasing. If node v has eccentricity t and for some node u, $d(u,v) = t$, then u is called an *eccentric node* of v. The characterization of r-decreasing graphs follows.

Theorem 5.5 For a graph G, the following statements are equivalent.

1. G is r-decreasing.
2. Each node $v \in V(G)$ has exactly one eccentric node $\bar{v}$ such that $e(v) = e(\bar{v})$.
3. There exists a decomposition of $V(G)$ into pairs $v, \bar{v}$ such that $d(v,\bar{v}) = r(G) > \max\{d(u,v), d(u,\bar{v})\}$ for all $u \in V(G) - \{v, \bar{v}\}$. □

Akiyama, Ando, and Avis [AAA2] defined the *eccentric graph* G_e of G to be the graph with $V(G_e) = V(G)$, where two nodes u and v of $V(G_e)$ are adjacent if one of them is an eccentric node of the other. They showed that a graph G on $2n$ nodes is r-decreasing if and only if $G_e = nK_2$, thereby rediscovering part of Gliviak's characterization. Recall from Chapter 2 that a graph G is a *unique eccentric node graph* if each node in G has precisely one eccentric node. Nandakumar [N1] showed that a unique eccentric node graph G is self-centered if and only if $G_e = nK_2$. This give the following.

Corollary 5.5 A graph G on $2n$ nodes is r-decreasing if and only if G is a self-centered unique eccentric node graph. □

EXERCISES 5.1

1. Every nontrivial self-complementary graph G has radius $r(G) = 2$.
2. Let u and v be nodes in G with $r(G) \geq 3$. Then
   $\deg u + \deg v \leq p - 2r(G) + 4$.
3. For an r-critical graph G, $r(G - v) < r(G)$ if and only if v is a peripheral node.
4. If G and H are r-decreasing, then the cartesian product $G \times H$ is also r-decreasing.

5. Let G have $p \geq 3$ nodes, and let $r \geq 3$ and $k \geq 2p + 2$ be given integers. Then there exists a k-regular, r-decreasing, radius-maximal graph Q with radius r which has G as an induced subgraph.

(Gliviak [G7])

6. Graph $G_e = nK_2$ if and only if G is r-critical.   (Nandakumar [N1])

7. Show that the graph in Figure 5.4 is not r-critical.

8. If G is an r-critical graph that is not nonseparable, then every peripheral node of G has degree 1.   (Gliviak [G7])

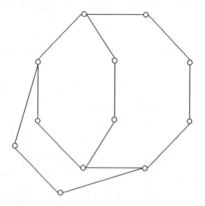

Figure 5.4   A self-centered graph that is not radius-critical.

5.2 SMALL DIAMETER

Moon and Moser [MM1] were first to show that almost all graphs have diameter two. Their result was generalized by Klee and Larman [KL1] and independently by Bollobás [B16] who obtained the most general result. Thus a discussion of graphs of small diameter includes most graphs. We focus here on extremal results for graphs having diameter two or three. Extremal results for general values of d will be presented in the next section.

Moore Graphs

One of the earlier results for graphs with small diameter concerns a special class of graphs called Moore graphs. The *distance degree sequence of a node v* is $dds(v) = (d_0(v), d_1(v), \ldots, d_{e(v)}(v))$, where $d_i(v)$ is the number of nodes at distance i from v. If every node of G has the same distance degree sequence, then G is said to be *distance degree regular (DDR)*. We shall examine these graphs in §9.2. A (k,t)-*Moore graph* is a k-regular graph G of diameter t where for some peripheral node $v \in G$, $d_i(v) = k \cdot (k-1)^{i-1}$ for each $i \geq 1$. A Moore graph is a DDR graph, so it is self-centered and self-median. Hence any node could play the role of the peripheral node v. The *girth* of a graph is the length of any shortest cycle. A (k,t)-Moore graph could be described as a k-regular graph with girth $g = 2t+1$ having the minimum possible number of nodes. These are a special example of a class of highly symmetric graphs called cages that we discuss in §8.2. Hoffman and Singleton [HS1] classified Moore graphs with diameters 2 and 3. They showed that the only Moore graph of diameter 3 is C_7. For diameter 2, the situation is far more complicated. They showed that for diameter 2 there are Moore graphs of degree 2, 3, 7, and possibly 57; and the $(2,2)$, $(3,2)$, and $(7,2)$-Moore graphs are unique. The $(2,2)$-Moore graph is C_5, and the $(3,2)$-Moore graph is the Petersen graph. The $(7,2)$-Moore graph (also called the $(7,5)$-cage or the *Hoffman-Singleton graph*) has 50 nodes. A description of its construction can be found in Biggs [B8,p.163]. The question as to whether the $(57,2)$-Moore graph actually exists remains open, which is not surprising since this graph would have 3250 nodes and 92,625 edges, yet a result of Aschbacher [A3] shows that it cannot be distance-transitive (see §8.3).

Diameter-Minimal and Diameter-Critical Graphs

Deleting an edge from a graph may cause its diameter to increase or stay the same, but it cannot decrease. A graph G is *diameter-minimal* if for all edges $e \in G$, $d(G - e) > d(G)$. Any edge that can be removed from G without affecting the diameter is called *superfluous*. Thus diameter-minimal graphs are the graphs with no superfluous edges. Suppose that G has diameter 2. Then every superfluous edge $e = uv$ is contained in a triangle, since otherwise removal of e would make $d(G) \geq d(u,v) \geq 3$. Gliviak [G6] established a number of existence results diameter-minimal graphs of diameter 2, which he refers to as graphs of *class B*. He partitions his study in terms of the types of induced cycles the graphs contain. Since a graph in B cannot contain an induced C_n for $n \geq 6$, there were $2^3 = 8$ types of graphs he had to consider. Note that B graphs having an induced C_5 but no induced C_4 or C_3 are precisely the Moore graphs of diameter 2. One of the more interesting results of [G6] is the following.

Theorem 5.6   Every graph G can be imbedded as an induced subgraph in a diameter-minimal graph of diameter 2.

Proof   Label the nodes of G by $v_1, v_2, \ldots, v_p$. Next, add new nodes $w, x, u_1, u_2, \ldots, u_{p+1}$, and edges $v_i u_i$, $w v_i$, $w u_{p+1}$, and $x u_i$. Finally, for each pair of distinct nonadjacent nodes v_i, v_j insert the edge $u_i u_j$. It is easy to verify that the resulting graph is diameter-minimal, has diameter 2, and has G as an induced subgraph.   □

Note that the graph constructed in Theorem 5.6 also has the property that distinct nodes have distinct neighborhoods.

By removing nodes rather than edges from G, the diameter could increase, decrease or stay the same. A graph G is *diameter-critical* if $d(G - v) \neq d(G)$ for every node $v \in G$. Note that in a diameter-critical graph G, some nodes of G may cause the diameter to increase when removed while others cause it to decrease. For example, if G consists of C_7 with a endedge attached, then removal of the node of degree one would cause the diameter to decrease, whereas the removal of any other node increases the diameter. One may then wonder whether a diameter-critical graph can have the property that the removal of each of its nodes causes the diameter to decrease or the removal of each of its nodes causes the diameter to increase. Gliviak [G9] answered this question by showing that a graph can have at most two nodes that decrease the diameter.

Theorem 5.7   Let G be a connected graph and let X be the set of all nodes x for which $d(G - x) < d(G)$. Then $|X| \leq 2$ and $d(G) - 2 \leq d(G - X) \leq d(G) - 1$ for $X \neq \emptyset$.

Proof   If $x \in X$ then $d_G(u, v) \leq d(G - x) < d(G)$ for all $u, v \in G - x$. Hence if $d(y, w) = d(G)$ for some pair of nodes $y, w \in G$, then y or w must be x. So each node in X is peripheral. Any pair of nodes $x, x' \in X$ must be antipodal nodes or else removing one of them could not decrease the diameter. But then if X contains three or more nodes, removal of one of them, say x^* would still leave an antipodal pair of nodes at distance $d(G)$ from one another in $G - x^*$, contradicting the fact that $d(G - x^*) < d(G)$. Thus $|X| \leq 2$.

By deleting a single node from G, the length of any longest path will decrease by at most one. And by deleting two nodes, the diameter may decrease by at most two. Hence the stated inequalities on the diameter hold.   □

A graph G is *d-increasing* if $d(G - v) > d(G)$ for all $v \in G$ and *d-decreasing* if $d(G - v) < d(G)$ for all $v \in G$. Theorem 5.7 implies that the only d-decreasing graphs are K_2 and $\overline{K}_2$. In [G9], Gliviak gave various results on embedding graphs into d-increasing graphs and into diameter-critical graphs that are not d-increasing as well as the following result for small diameter.

Theorem 5.8 Let G be a diameter-critical graph of diameter $d \leq 3$ that is not d-increasing. Then G is a path of length d.   □

Trees of Small Diameter

Although there are infinitely many trees of diameter at most three, they are easy to describe. They are the graphs K_1, K_2, the stars $K_{1,n}$, and the double stars $S_{m,n} = \overline{K}_m + K_1 + K_1 + \overline{K}_n$. These trees have been used in various situations in the literature, perhaps the most interesting of which is in decomposition problems or analogously, in packing problems.

The slight difference between packing problems and decomposition problems is one of generality. In a typical packing problem, one begins with a well-known class of graphs such as the K_p or $K_{m,n}$ and wants to determine whether one can color the edges of the graph so that the color classes determine some fixed set of trees of small diameter. The following interesting result of this genre is due to Bourgeois, Hobbs, and Kasiraj [BHK1].

Theorem 5.9 Suppose $T_2, T_3, \ldots, T_p$ are trees such that T_i has order i and $d(T_i) \leq 3$. Then $T_2, T_3, \ldots, T_p$ can be packed into K_p.   □

An example providing a packing of trees T_i into K_5 is given in Figure 5.5. Bourgeois, Hobbs and Kasiraj also showed that the result in Theorem 5.9 is still true if one allows at most one of the trees to have diameter greater than three.

In decomposition problems, one generally begins with an *arbitrary* graph with the goal of coloring the edges so that each color class corresponds to some *arbitrary* tree within some fixed class of trees. For example, Lovász [L7] showed that any graph G whose maximum matching has size m can be decomposed into m double stars and $p - 2m$ stars.

Minimal Blocks of Small Diameter

Recall that a graph is nonseparable if it is connected and has no cutnodes. Such a graph has only one block, and for that reason, the graph itself is

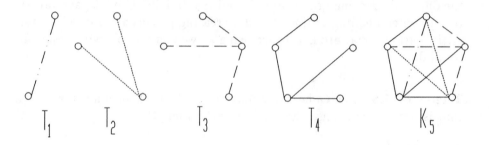

Figure 5.5 A packing of trees into K_5.

often called a *block*. A block G is minimal if $G - e$ is not a block for each edge $e \in G$. These graph were characterized independently by Plummer [P7] and Dirac [D9]. Recently, Harary and Tindell [HT2] considered restricting the diameter of the blocks and characterized the minimal blocks of diameter 2 and 3.

Theorem 5.10 The minimal blocks of diameter 2 are as follows:

1. $K_{2,p-2}$ with $p \geq 4$, and
2. The graph formed from the double star $S_{m,n}$ by adding a new node v and joining v to each endnode of $S_{m,n}$.   □

The minimal blocks of diameter three are far more complicated to describe. We refer to [HT2] for further details.

EXERCISES 5.2

1. The only unique eccentric node graphs of diameter two are the graphs $K_{2n} - nK_2$.
2. If $d(G) = 2$, $\Delta(G) = k$, and $p = k^2$, then G is a 4-cycle C_4.
                        (Erdös, Fajtlowicz, and Hoffman [EFH1])
3. The only diameter-minimal graphs of diameter two with no induced C_4 or C_5 are the stars $K_{1,n}$.   (Gliviak [G6])
4. In a diameter-minimal graph there is at most one block containing a cycle.   (Plesník [P3])
5. If $\delta(G) \geq 2$ and $g(G) \geq d(G) + 3$, then G is diameter-critical.

6. If $d(G) = 2$ and $\Delta(G) = p - 2$, then $|E(G)| \geq 2p - 4$.

7. If G is diameter-critical with $d(G) = 2$ and $\kappa(G) = 1$, then $G = P_3$.

8. The *double m-star* is the double star $S_{m,m}$. Every $(2m + 1)$-regular graph $(m \geq 1)$ with a perfect matching is decomposable into double m-stars.   (Dean [D1])

5.3   DIAMETER

There are a huge number of extremal results concerning graphs of diameter d. Many of those theorems deal with random graphs and are asymptotic in nature. To avoid a 100 page section, we shall focus only on those observations on random graphs which are crucial to the study of other distance problems. For additional discussion concerning random graphs, see the books by Bollobás [B15] or Palmer [P1] or the survey article by Bermond and Bollobás [BB1].

Diameter-Critical and Diameter-Minimal Graphs

In the last section we presented results on diameter-critical graphs of small diameter. We now complete that discussion by looking at arbitrary values of $d(G)$. These results are also due to Gliviak [G9].

Theorem 5.11   For any graph G and integer $k \geq 4$, there exists a diameter-critical graph H with diameter k containing G as an induced subgraph.

Proof   Begin by adding an additional node $v_{1,0}$ to G and joining $v_{1,0}$ to each node of G to form graph F. Label the nodes of G by $v_{1,t}$, $1 \leq t \leq p$. Next take a copy of $\overline{F}$ with nodes $v_{k-1,t}$, $0 \leq t \leq p$. Add additional nodes $v_{i,t}$, $2 \leq i \leq k-2$, $0 \leq t \leq p$, and join each pair of nodes $v_{1,t}$ and $v_{k-1,t}$ by the path $v_{1,t}, v_{2,t}, \ldots, v_{k-1,t}$. Finally, add two more nodes u and w, join u to w and to all nodes $v_{k-1,t}$. (See Figure 5.6.)

It is easy to check that the resulting graph H has diameter d and it clearly contains G as an induced subgraph. It remains only to verify that H is diameter-critical. Graph $H - w$ has diameter $d-1$ and $d(H - u) = \infty$. Upon removing some node $v_{i,j}$ with $2 \leq i \leq k - 1$, one finds the distance $d(w, v_{1,j}) > k$ in $G - v_{i,j}$, so $d(G - v_{i,j}) > k$. Removing $v_{1,j}$ yields the distance $d(v_{2,j}, x) > k$ in $G - v_{1,j}$ for $x \in N_F(v_{1,j})$. Hence, each node's removal alters the diameter, so H is diameter-critical.   □

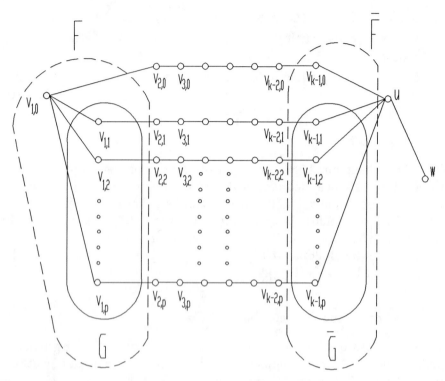

Figure 5.6   A diameter-critical graph containing a given graph.

Bosák, Rosa, and Znám [BRZ1] found the following bound on the maximum degree of a node in a graph G in terms of the diameter:

$$\Delta(G) \le p - d(G) + 1.$$

For diameter-critical graphs, Gliviak [G9] found a bound for the *minimum* degree in terms of diameter.

Theorem 5.12   Let $G \ne K_2$ be a diameter-critical graph on p nodes. Then $\delta(G) \le \lfloor (p - d + 1)/2 \rfloor$.   □

Ore [O3] discovered the following bound on the number of edges in a graph.

Theorem 5.13   For any (p, q)-graph of diameter d, we have

$$q \le d + \tfrac{1}{2}(p - d - 1)(p - d + 4).$$ □

Brigham and Dutton [BD1] developed an interactive computer program that allows one to specify values of various graphical invariants in order to query for possible values of other invariants. The database for their program consists of a huge number of theorems involving graphical equalities and inequalities. By using that system they found several new relationships between invariants.

Geodetic Connectivity

A connected noncomplete graph G is n-*geodetically connected* if the removal of at least n nodes are required to increase the distance between every pair of nonadjacent nodes. The *geodetic connectivity* is the maximum n such that G is n-geodetically connected. If G is n-geodetically connected, then it is obviously n-connected, but the converse is not true. For example, the graph $K_4 + \overline{K}_2$ is 3-connected, but only 2-geodetically connected. Note that every graph with geodetic connectivity equal to one is diameter-critical, but the converse is not true even if diameter-critical is replaced by d-increasing. Entringer, Jackson, and Slater [EJS2] characterized n-geodetically connected graphs.

Theorem 5.14   The following assertions are equivalent for a graph G:

1. G is n-geodetically connected.
2. G is connected and every two nodes at distance two from one another are joined by at least n geodesics.
3. For every pair of distinct nonadjacent nodes u and v, any set of $m \le n$ disjoint u-v geodesics is contained in a set of n disjoint u-v geodesics.
4. For any $n + 1$ distinct nodes $v_0, v_1, v_2, \ldots, v_n$, G contains disjoint v_0-v_1, v_0-v_2, $\ldots$, v_0-v_n geodesics. $\square$

Diameter and Connectivity

Among all invariants studied in connection with diameter extremal problems except maximum degree, connectivity has played the greatest role. It has generated the most interest in problems involving regular graphs. A minimum (t, k, n)-graph is an n-regular graph G of minimum order with $\kappa(G) = k$ and diameter $d(G) = t$. This class of graphs was introduced by Klee and Quaif [KQ2] for cubic graphs only. Myers [M14,15] examined minimum $(t, 3, 3)$-graphs, that is, minimum order 3-connected cubic graphs with diameter t. Bhattacharya [B6] studied the more general problem of n-connected, n-regular graphs with diameter t, and determined the minimum order for such graphs.

Theorem 5.15   The minimum order p of an n-regular, n-connected graph of diameter t is

$$\begin{cases} n+1 & t = 1, \\ n(t-1)+3 & n \text{ odd and } t \text{ even, and} \\ n(t-1)+2 & \text{otherwise.} \end{cases} \qquad \square$$

In §2.1, we described Ore's characterization of diameter-maximal graphs. Ore [O3] also characterized the diameter-maximal graphs of connectivity n having the maximum number of edges among all such graphs of order p. Their structure is similar to that of diameter-maximal graphs.

Theorem 5.16   A diameter-maximal graph G of diameter $t \geq 4$, connectivity n and order p having the maximum number of edges has the form

$$K_1 + K_n + K_{a_2} + K_{a_3} + \cdots + K_{a_{t-2}} + K_n + K_1$$

with $a_i = n$ for each i except possibly one or two consecutive a_i for which $a_i > n$. $\qquad \square$

Caccetta and Smyth [CS1] showed that the structure for diameter-maximal graphs with edge-connectivity n, diameter t, order p and the maximum number of edges is similar, yet somewhat more complicated.

Theorem 5.17   A diameter-maximal graph G of diameter $t \geq 6$, edge-connectivity n and order p having the maximum number of edges has the form

$$K_1 + K_n + K_{a_2} + K_{a_3} + \cdots + K_{a_{t-2}} + K_n + K_1,$$

where every triple (a_{i-1}, a_i, a_{i+1}), $3 \leq i \leq t-3$, except possibly one, contains exactly $n+1$ nodes. The exceptional triple is either (a_2, a_3, a_4) or $(a_{t-4}, a_{t-3}, a_{t-2})$. $\qquad \square$

Some Random Results

The random graph model $\Gamma(n, p)$ consists of all graphs with node set $V(G) = \{v_1, v_2, \ldots, v_n\}$ whose edges are chosen independently with probability p. Hence, if G is a graph with node set V and has m edges, then the probability of G is $\Pr(\{G\}) = p^m(1-p)^{N-m}$, where $N = \binom{n}{2}$. Let *lg x* mean $\log_2 x$. For a property $\mathcal{P}$, we say that a random graph $G \in \Gamma(n,p)$ has property $\mathcal{P}$ *almost surely (a.s.)* if $\Pr(G$ has property $\mathcal{P}) \longrightarrow 1$ as $n \longrightarrow \infty$. Perhaps the most useful results concerning diameter of random graphs are the following due to Bollobás [B16].

Theorem 5.18

1. If $p^2 n - 2 \lg n \longrightarrow \infty$ and $n^2(1-p) \longrightarrow \infty$, then $G \in \Gamma(n,p)$ almost surely has diameter 2.

2. If the functions $d = d(n) \geq 3$ and $0 < p = p(n) < 1$ satisfy the three conditions

    a. $(1/d)\lg n - 3\lg\lg n \longrightarrow \infty$,

    b. $p^d n^{d-1} - 2\lg n \longrightarrow \infty$,   and

    c. $p^{d-1} n^{d-2} - 2\lg n \longrightarrow -\infty$

    then $G \in \Gamma(n,p)$ almost surely has diameter $d(n)$.   □

It is easy to see that with $p = 1/2$ both expressions in part 1 of Theorem 5.18 approach ∞ as $n \longrightarrow \infty$. Because of this one can say that almost all graphs have diameter 2. In fact, another result of Bollobás says that for a very large range of p (including $p = 1/2$) almost no graph has a node of full degree. Hence almost no graph has radius one. Since $r(G) \leq d(G)$, almost all graphs have $r(G) = d(G) = 2$, that is, almost all graphs are self-centered with diameter 2.

Buckley and Palka [BP2] used some powerful results of Burtin [B30,31] to examine the central and peripheral ratios of random graphs as well as the asymptotic distributions of the size of the center and the periphery of a random graph.

EXERCISES 5.3

1. Construct a minimum order 4-regular, 4-connected graph of diameter 3.   (Bhattacharya [B6])

2. Let $K(G)$ be the clique graph of G. Then
    $d(G) - 1 \leq d(K(G)) \leq d(G) + 1$.   (Hedman [H19])

3. An edge $e = uv$ is a *dominating edge* in graph G if every node of G is adjacent to at least one of u and v. Almost no graph contains a dominating edge.

5.4   LONG PATHS AND LONG CYCLES

In Chapter 4, we discussed hamiltonian and nearly hamiltonian graphs. The subject of this section is related to some of those hamiltonicity questions. We consider detours in graphs, long paths that avoid certain nodes of the the graph, and conditions that guarantee a path or cycle of some given length.

Paths and Trails

Recall that a *trail* is a walk in which no edge appears more than once. Thus in a trail, nodes can be revisited, but edges cannot. In a path, neither nodes nor edges may be repeated. The *trail number* $tr(G)$ of a graph G is the maximum length of a trail in it. This invariant was examined by Bollobás and Harary [BH1] who determined exactly the maximum trail number among all (p, q)-graphs.

Theorem 5.19   The maximum trail number among all graphs on p nodes and q edges is

$$\begin{cases} q & p \text{ odd or } q \leq \binom{n}{2} - \frac{n}{2} + 1; \\ \binom{n}{2} - \frac{n}{2} + 1 & \text{otherwise.} \end{cases}$$

Proof   If p is odd then K_p is eulerian and the p nodes of K_p along with the first q edges of any eulerian trail of K_p form a (p, q)-graph G with $tr(G) = q$. If p is even, then $K_p - (\frac{p}{2} - 1)K_2$ has a spanning trail that includes every edge. Hence for $q \leq \binom{n}{2} - \frac{n}{2} + 1$, the first q edges in such a trail again produces a (p, q)-graph G with $tr(G) = q$. On the other hand, if p is even, $tr(G)$ can be no larger than $\binom{n}{2} - \frac{n}{2} + 1$ since any subgraph H formed by the edges of a trail in G has at most two nodes of degree $p - 1$. So

$$tr(G) = |E(H)| \leq \tfrac{1}{2}\left(2(n-1) + (n-2)^2\right) = \binom{n}{2} - \frac{n}{2} + 1. \qquad \square$$

In [BH1], close upper bounds for the minimum trail number among all (p, q)-graphs were also obtained.

When studying long paths in graphs many surprising results are found. For example, there are graphs with the property that none of their diametral paths contain a central node. Buckley and Lewinter [BL3] determined the range of the diameters for such graphs.

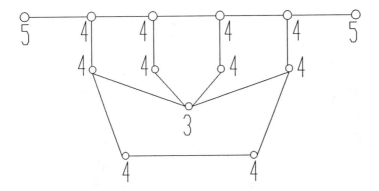

Figure 5.7   A graph with all diametral paths avoiding the center.

Theorem 5.20   Suppose that all diametral paths of G avoid the center. Then

$$r(G) + 2 \leq d(G) \leq 2r(G) - 1$$

and all pairs of values in the given range are attainable.   □

Note that the constraints on diameter for such a graph imply that a smallest graph with every diametral path avoiding the center has radius 3 and diameter 5 as shown in Figure 5.7.

Buckley and Harary [BH5] determined the maximum length of a longest induced path among all graphs with p nodes and q edges.

Theorem 5.21   Among all (p, q)-graphs G, the maximum length of a longest induced path is given by:

$$\begin{cases} q & q < p; \\ \max\{t : t(t-1) \leq p(p-1) - 2q\} + 1 & q \geq p. \end{cases}$$ □

In [BH5], the authors also determined the maximum length of a longest induced path among all (p, q) bipartite graphs.

Detours

A *detour* in a graph G is a path of maximum length and the length of such a path is called the *detour number* $dn(G)$. This concept was studied by Kapoor, Kronk, and Lick [KKL1] who found relations between $dn(G)$ and other invariants. Among their results is the following.

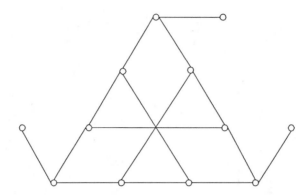

Figure 5.8   A graph with only 12 nodes having a detour avoiding each given node.

Theorem 5.22   If G is a connected graph on p nodes then

$$dn(G) \geq \min\{p - 1, 2\delta(G)\}. \qquad\qquad \square$$

A surprising result concerning detours is that there are graphs in which for each node, there is a detour which avoids it. H. Walther found the first such example which had 25 nodes thereby answering a question of T. Gallai. Zamfirescu [Z2] cut Walther's bound in half when he found a graph on 12 nodes which also has the property. His graph is displayed in Figure 5.8.

Even more surprising is that one can require that G be 2-connected, or even 3-connected and planar and still have graphs with detours avoiding any given node (or even any given pair of nodes. For a lively discussion of this and related problems, see [Z3]. The *circumference* of a graph is the length of its longest cycle, and a cycle achieving that length is a *circumcycle*. Circumcycles avoiding a given node are also discussed in [Z3].

Long Cycles and Toughness

Chvátal [C15] introduced the concept of toughness to study hamiltonicity. Let $\omega(H)$ denote the number of components in a graph H. A graph G is *t-tough* if each subset $S \subset V(G)$ with $\omega(G - S) > 1$ satisfies

$$|S|/\omega(G - S) \geq t.$$

The toughness of a graph is the maximum value of t for which it is t-tough. A necessary condition for a graph to be hamiltonian is that it be 1-tough.

Chvátal found 3/2-tough nonhamiltonian graphs, but conjectured that there is some real number t^* such that all t^*-tough graphs are hamiltonian. The conjecture remains open.

Toughness has been used recently by a number of authors not only to consider hamiltonicity, but also to bound the circumference in a graph that might not be hamiltonian. For example, Bauer and Schmeichel [BS1] proved the following which had been conjectured by A. Ainouche and N. Christofides.

Theorem 5.23 If G is 1-tough with order $p \geq 3$ such that $\deg u + \deg v \geq k$ for all distinct nonadjacent nodes u, v, then the circumference of G is at least $\min\{p, k+2\}$. □

Bauer, Morgana, Schmeichel and Veldman [BMSV1] extended this by considering triples rather than pairs of nodes.

Theorem 5.24 Suppose that G is 1-tough with order $p \geq 3$ such that $\deg u + \deg v + \deg w \geq k$ for all independent triples of nodes u, v, w. Then the circumference of G is at least $\min\{p, p/2 + k/3\}$. □

In some studies on longest cycles, 2-connectedness is assumed rather than a toughness constraint. An example of this is the following result of Fan [F1].

Theorem 5.25 Let G be 2-connected with order p an let k be an integer with $3 \leq k \leq p$. If for all pairs of nodes u, v at distance two from one another $\max\{\deg u, \deg v\} \geq k/2$, then the circumference of G is at least k. □

Further results of this flavor as well as additional references are given in [BSV1] and [BMSV1].

Cycles of a Given Length

In some instances, authors have been concerned with whether a graph contains cycles of some given length. Of particular interest has been the case of graphs containing a cycle of some residue class modulo b. Along these lines, Bollobás [B14] answered a question of S. Burr and P. Erdös with the following result.

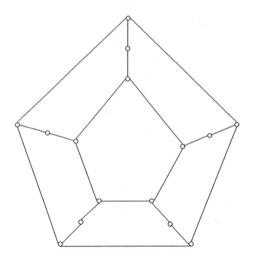

Figure 5.9   A graph with circumcycles avoiding each node.

Theorem 5.26   For a graph G, if $k \geq 3$ is an odd positive integer and let

$$\delta(G) \geq \frac{(k+1)^k - 1}{k} \quad \text{or} \quad q \geq \frac{(k+1)^k - k - 1}{k}p$$

then for every natural number t, G contains a cycle of length t mod k.   □

As a particular example, this result asserts that if $\delta(G) \geq 21$, then G contains a cycle of length 0 mod 3. Barefoot, Clark, Douthett, Entringer, and Fellows [BCDEF1] set out to determine the actual structure of such graphs. They showed that all cubic graphs, subdivisions of 3-connected cubic graphs on $p \geq 10$ nodes, and subdivisions of graphs with at least $3p - 5$ edges always contain a cycle of length 0 mod 3. Furthermore, they showed the bound $3p - 5$ is sharp.

Very recently, Erdős, Faudree, Gyárfás and Schelp [EFGS1] obtained the following result which determines the length of a cycle of a given size in a nonbipartite graph in terms of the minimum degree.

Theorem 5.27   Let $k \geq 3$ be a fixed positive integer. If G is a 2-connected nonbipartite graph on p nodes with $\delta(G) \geq 2p/(k+2)$, then for p large (as a function of k) either G contains the cycle C_k or G is isomorphic to the graph obtained from C_{k+2} by replacing each of its nodes by an independent set of order $p/(k+2)$.   □

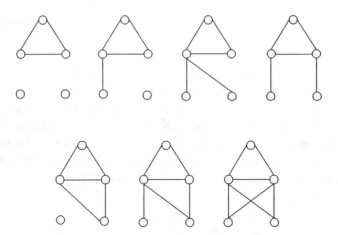

Figure 5.10   Induced subgraphs of G with $g(G) = g(\overline{G}) = 3$.

EXERCISES 5.4

1. Determine all graphs that contain P_4 but not P_5.

2. The graph G in Figure 5.9 has the property that for each node $v \in G$, it has a circumcycle that avoids v.   (C. Thomassen)

3. Determine the toughness of the wheels $W_{1,n}$.

4. Find a subdivision of K_4 with no cycle of length 0 mod 3. What is the smallest number of nodes?

5. Determine the detour number and circumference of each graph in the figures of Chapter 5.

6. Let $\mathfrak{F}$ be the set of all graph G for which $g(G) = g(\overline{G}) = 3$. Then $G \in \mathfrak{F}$ if and only if G contains one of the graphs in Figure 5.10 as an induced subgraph.   (Akiyama and Harary [AH1])

FURTHER RESULTS

Peyrat, Rall, and Slater [PRS1] answered a question of Hedman [H19] concerning the diameters of clique graphs. Let G be a graph with diameter $d(G) = n$. Then they showed its clique graph $K(G)$ has diameter $n + 1$ if and only if G has cliques C and D such that $d(x, y) = n$ for every pair of nodes $x \in C$ and $y \in D$.

A graph G is *diameter edge-invariant (d.e.i)* if its diameter is unchanged by the deletion of an edge, that is, $d(G - e) = d(G)$ for all $e \in G$. These graphs were studied by Lee [L2]. A *critical diameter edge-invariant graph* is a d.e.i. graph G such that for all $v \in G$, $G - v$ is not d.e.i. Lee and Wang [LW1] develop constructions of these graphs and examine their properties.

For and graph G, let $ch^+(G)$ be the least number of edges whose addition to G decreases the diameter, and let $ch^-(G)$ be the least number of edges whose deletion increases the diameter. Similarly, let $un^+(G)$ be the maximum number of edges that may be added to G without changing the diameter, and let $un^-(G)$ be the greatest number of edges whose removal does not affect the diameter. Graham and Harary [GH2] examine these invariants for the hypercubes. In particular, they show

$$\lim_{n \longrightarrow \infty} \frac{un^-(Q_n)}{n2^{n-1}} = 1,$$

which means that almost all edges of Q_n may be removed without altering the diameter. They also determine that

$$un^+(G) = \binom{2n}{n-1} + \frac{1}{2}\binom{2n}{n} - (n+1)2^{n-1}.$$

CHAPTER 6

Matrices

A graph is completely determined either by its adjacencies or by its incidences. This information can be conveniently stated in matrix form. Indeed, with a given graph, adequately labeled, there are associated several matrices, including the adjacency matrix, incidence matrix, distance matrix, cycle matrix, and cocycle matrix. It is often possible to make use of these matrices in order to identify certain properties of a graph. The classic theorem on graphs and matrices is the Matrix-Tree Theorem, which gives the number of spanning trees in a given graph.

6.1 THE ADJACENCY MATRIX

Suppose we want to test a conjecture about graph with the aid of a computer. A standard technique is to represent, store, and manipulate the graph in computer memory using a matrix. A common matrix used in this way is the adjacency matrix. In a *binary matrix* each entry is 0 or 1. Let the nodes of G be labeled $v_1, v_2, \ldots, v_p$. The *adjacency matrix* $A = A(G) = [a_{ij}]$ of G is the binary matrix of order p

$$a_{ij} = \begin{cases} 1 & \text{if } v_i \text{ is adjacent with } v_j \\ 0 & \text{otherwise.} \end{cases}$$

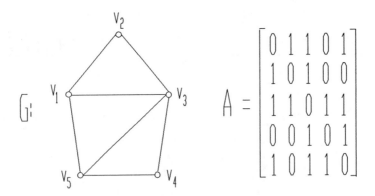

Figure 6.1 A labeled graph and its adjacency matrix.

Since graphs have no loops, the diagonal of A is zero. Thus there is an immediate one-to-one correspondence between labeled graphs on p nodes and $p \times p$ symmetric binary matrices with zero diagonal.

Figure 6.1 shows a labeled graph G and its adjacency matrix A. One immediate observation is that the row sums of A are the degrees of the nodes in G. Thus the degree sequence can be obtained at once from the adjacency matrix.

Because of this correspondence between graphs and matrices, every graph theoretic concept is reflected in its adjacency matrix. For example, consider the statement that a graph G is connected if and only if there is no partition $V = V_1 \cup V_2$ of its nodes such that no edge joins a node of V_1 with a node of V_2. In matrix terms, we may say that G is connected if and only if there is no labeling of its nodes such that its adjacency matrix has the form

$$A = \begin{pmatrix} A_{11} & 0 \\ 0 & A_{22} \end{pmatrix}$$

where A_{11} and A_{22} are square.

A graph G does not always come conveniently labeled. It may have labels other than $v_1, v_2, \ldots, v_p$ or G may have no labels. In either case, we can arbitrarily assign the labels $v_1, v_2, \ldots, v_p$ to the nodes of G. Different label assignments will produce different matrices. There is nothing to worry about; all information about a graph G is contained in the matrix A no matter how we label the graph. If A_1 and A_2 are adjacency matrices which arise from two different labelings of the same graph G, then for some permutation matrix P, $A_1 = P^{-1}A_2P$. Sometimes a labeling is irrelevant, as in the following results which interpret the entries of the powers of the adjacency matrix of a digraph D, which of course apply as well to a graph G.

Theorem 6.1   Let D be a labeled digraph with adjacency matrix A. Then the (i,j)-entry of A^n is the number of walks of length n from v_i to v_j.

Proof   The proof is by induction on n. The result is obvious for $n = 1$ since $a_{ij}^{(1)} = a_{ij} = 1$ if and only if the arc $v_i v_j$ is present. Now assume that $a_{ij}^{(n-1)}$ is the number of distinct walks of length $n - 1$ from v_i to v_j. Since $A^n = A^{n-1} A$, its entries $a_{ij}^{(n)}$ are found by

$$(6.1) \qquad a_{ij}^{(n)} = \sum_{k=1}^{p} a_{ik}^{(n-1)} a_{kj}.$$

Since every walk of length n from v_i to v_j consists of a walk of length $n - 1$ from v_i to some node v_k followed by the arc $v_k v_j$, the inductive hypothesis and (6.1) yields the desired result. $\square$

Corollary 6.1a   In a digraph D, the off-diagonal entry $a_{ij}^{(2)}$ of A^2 is the number of paths of length two from v_i to v_j. $\square$

Corollary 6.1b   In a graph G, the diagonal entry $a_{ii}^{(2)}$ of A^2 is the degree of v_i. $\square$

Corollary 6.1c   If a graph G is connected, the distance between v_i and v_j for $i \neq j$ is the least integer n for which $a_{ij}^{(n)} > 0$. $\square$

A simple yet useful observation about the adjacency matrix concerns bipartite graphs. A graph G is bipartite if and only if there is a labeling of its nodes so that $A(G)$ has the form

$$A(G) = \begin{pmatrix} 0 & B \\ B^t & 0 \end{pmatrix}.$$

Determinants and Inverses

Since the adjacency matrix of a graph is square, we can investigate its determinant. Clearly the determinant of A is independent of the labeling of the nodes of G. Hence we may say that the *determinant of a graph G is* the determinant of any adjacency matrix of G. If two nonadjacent nodes of a graph G have the same neighborhood, then $\det A = 0$. This follows from the fact that two rows of A would be identical. Thus a complete bipartite graph $K_{m,n}$ has zero determinant except when $m = n = 1$.

An *ordinary linear subgraph* (*o'graph* for short as in [BDH1]) of G is a spanning subgraph whose components are single edges (K_2) or cycles (C_n). A component is *even* if it has an even number of nodes. Thus each K_2 component of an o'graph is even. Let ε be the number of even components and let c be the number of cycles in an o'graph. Harary [H11] determined a formula for the determinant of a graph.

Theorem 6.2   Let A be any adjacency matrix of a graph G. Then the determinant of G is given by

$$(6.2) \qquad\qquad \det A = \sum (-1)^\varepsilon \cdot 2^c,$$

where the sum is taken over all o'graphs of G.   □

Corollary 6.2   For a graph G, $|\det A| = 1$ implies that G has a 1-factor.

Proof   If G has no o'graph, then $\det G = 0$ as there are no terms in the expansion of (6.2). Thus, suppose G has an o'graph. If every o'graph contains a cycle, then every term from the sum in (6.2) is even because it has $c > 0$. Thus if $|\det A| = 1$, then some o'graph must be acyclic. Hence, $|\det A| = 1$ implies that G has a 1-factor.   □

The graph $K_{n,n}$, $n > 1$, shows that the converse to Corollary 6.2 does not hold. Harary and Minc [HM5] defined a graph G to be *invertible* if A^{-1} is the adjacency matrix of some graph. Since an integral matrix M (with all entries integers) has an integral inverse if and only if $\det M = \pm 1$, invertible graphs satisfy $|\det A| = 1$.

Theorem 6.3   The only invertible graphs are the matchings nK_2.   □

A *signed graph* is a graph with each edge labeled by a sign ($+$ or $-$). A signed graph is *balanced* if every cycle contains an even number of negative edges. This concept has proved useful in social science applications and in studying the stability of political alliances [H10]. The structure of balanced signed graphs was determined in [H4].

In a signed graph, the positive edges are drawn solid and the negative edges are drawn dashed. An adjacency matrix of a signed graph is symmetric, has zeros on the diagonal, and each entry is 0, 1, or -1. A *signed invertible graph* (*s-invertible* for short) is a graph G for which A^{-1}

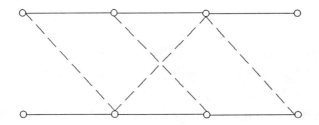

Figure 6.2   The signed inverse of the s-invertible graph P_8.

is the adjacency matrix of some signed graph H. In this case, we call H the *signed inverse* of G and write $G^{-1} = H$. The signed inverse of P_8 is shown in Figure 6.2.

Just as a graph G must have $\det A = \pm 1$ to have an inverse, this condition is necessary for the existence of a signed inverse because a signed graph also has an integral adjacency matrix. But this condition is not sufficient. For example, the graph in Figure 6.3 has determinant -1 but is not s-invertible.

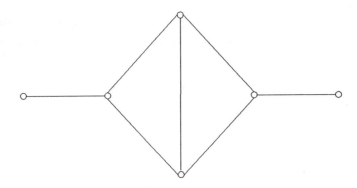

Figure 6.3   A graph G with $|\det A| = 1$ which is not s-invertible.

Buckley, Doty, and Harary [BDH1] characterized s-invertible trees.

Theorem 6.4   A tree T is s-invertible if and only if it has a perfect matching.   □

Note that "tree" in Theorem 6.4 cannot be replaced "graph" as Figure 6.3 would be a counterexample.

EXERCISES 6.1

1. Find the determinants of P_n, W_n, and K_n.

2. The path P_n is s-invertible if and only if n is even.

3. The signed inverse $(P_{2k})^{-1}$ is balanced.

4. A graph G is bipartite if and only if for all odd n, every diagonal entry of A^n is 0.

5. The inner product of any two rows of A is at most one if and only if G contains no C_4.

6. For a graph G, the diagonal entry $a_{ii}^{(3)}$ of A^3 is twice the number of triangles containing v_i.

7. The *trace* $tr(M)$ of a square matrix M is the sum of its diagonal entries. Let T be the number of of triangles in a graph G with adjacency matrix A. Then $T = tr(A^3)/6$.

8. For a signed graph S let S^* be its underlying graph (with the signs removed). For any graph G, the corona $G \circ K_1$ is s-invertible. Furthermore, $((G \circ K_1)^{-1})^* \cong G \circ K_1$.

    (Buckley, Doty, and Harary [BDH1])

9. Let G be a connected graph with adjacency matrix A. What can be said about A if

    a. v_i is a cutnode?

    b. $v_i v_j$ is a bridge?

10. Two graphs G_1 and G_2 are *cospectral* if the polynomials $\det(A_1 - tI)$ and $\det(A_2 - tI)$ are equal. There are just two different cospectral graphs with 5 nodes.

    (Harary, King, Mowshowitz, and Read [HKMR1])

11. An *eigenvalue* of G is a root of its *characteristic polynomial* $\det(A_1 - tI)$. A connected graph of order p and diameter d has at least $d + 1$ and at most p distinct eigenvalues.

12. Graph G is connected if and only if $(A + I)^{p-1}$ has no zero entries, i.e., $(A + I)^{p-1} > 0$.

6.2   THE INCIDENCE MATRIX

A second matrix associated with a graph G in which both its nodes and edges are labeled is the incidence matrix. Let the nodes of G be labeled

$v_1, v_2, \ldots, v_p$ and the edges be labeled $e_1, e_2, \ldots, e_q$. Then the *incidence matrix* B of G is the $p \times q$ binary matrix in which

$$b_{ij} = \begin{cases} 1 & \text{if } v_i \text{ is incident with } e_j \\ 0 & \text{otherwise.} \end{cases}$$

As with the adjacency matrix, the incidence matrix determines G up to isomorphism. In fact, any $p-1$ rows of B determine G since each row is the sum of all others modulo 2.

We now note several simple observations about B. Since each edge has two endnodes, each column of B contains exactly two 1s. Each 1 in row i corresponds to an edge incident with v_i. Thus, the number of 1s in row i is deg v_i. The next theorem relates the adjacency matrix of the line graph of G to the incidence matrix of G. Recall that B^t is the transpose of matrix B.

Theorem 6.5   For any (p,q)-graph G with incidence matrix B,
$A(L(G)) = B^t B - 2I_q$.   □

Let M denote the matrix obtained from $-A$ by replacing the ith diagonal entry by deg v_i. In the proof of our next theorem, we use the following algebraic result, called the Binet-Cauchy Theorem.

Lemma 6.6   If P and Q are $m \times n$ and $n \times m$ matrices, respectively, with $m \leq n$, then det PQ is the sum of the products of corresponding major determinants of P and Q.   □

In Lemma 6.6, a major determinant of P or Q has order m, and the phrase "corresponding major determinants" means that the columns of P in the one determinant are numbered like the rows of Q in the other. The following theorem is contained in the pioneering work of Kirchkoff [K2].

Theorem 6.6   (*Matrix-Tree Theorem*) Let G be a connected labeled graph with adjacency matrix A. Then all cofactors of the matrix M are equal and their common value is the number of spanning trees of G.

Proof   We begin the proof by changing either of the two 1s in each column of the incidence matrix B of G to -1, thereby forming a new matrix E. (This amounts to arbitrarily orienting the edges of G and taking E as the incidence matrix of this oriented graph.)

The i,j-entry of EE^t is $e_{i1}e_{j1} + e_{i2}e_{j2} + \cdots + e_{iq}e_{jq}$, which has the value deg v_i if $i = j$, -1 if v_i and v_j are adjacent, and 0 otherwise. Hence $EE^t = M$.

Consider any submatrix of E consisting of $p - 1$ of its columns. This $p \times (p - 1)$ matrix corresponds to a spanning subgraph H of G having $p - 1$ edges. Remove an arbitrary row, say the kth, from this matrix to obtain a square matrix F of order $p - 1$. We will show that $|\det F|$ is 1 or 0 according as H is or is not a tree. First, if H is not a tree, then because H has p nodes and $p - 1$ edges, it is disconnected, implying that there is a component not containing v_k. Since the rows corresponding to the nodes of this component are dependent, $\det F = 0$. On the other hand, suppose H is a tree. In this case, we can relabel its edges and nodes other than v_k as follows: Let $u_1 \neq v_k$ be an endnode of H, and let y_1 be the edge incident with it; let $u_2 \neq v_k$ be any endnode of $H - u_1$ and y_2 its incident edge, and so on. This relabeling of the nodes and edges of H determines a new matrix F' which can be obtained by permuting the rows and columns of F independently. Thus $|\det F'| = |\det F|$. However, F' is lower triangular with every diagonal entry $+1$ or -1; hence $|\det F| = 1$.

We apply Lemma 6.6 to calculate the first principal cofactor of M. Let E_1 be the $(p - 1) \times q$ submatrix obtained from E by striking out its first row. By letting $P = E_1$ and $Q = (E_1)^t$, we find from the lemma, that the first principal cofactor of M is the sum of the products of the corresponding major determinants of E_1 and $(E_1)^t$. Obviously, the corresponding major determinants have the same value. We have seen that their product is one if the columns from E_1 correspond to a spanning tree of G and is 0 otherwise. Thus the sum of these products is exactly the number of spanning trees.

The equality of all the cofactors, both principal and otherwise, holds for every matrix whose row sums and column sums are all zero, completing the proof.   □

To illustrate the Matrix-Tree Theorem, we consider a labeled graph G taken at random, say $K_4 - e$. This graph, shown in Figure 6.4, has eight spanning trees, since the $(2,3)$-cofactor, for example,

$$\text{of} \quad M = \begin{pmatrix} 3 & -1 & -1 & -1 \\ -1 & 2 & -1 & 0 \\ -1 & -1 & 3 & -1 \\ -1 & 0 & -1 & 2 \end{pmatrix} \quad \text{is} \quad -\begin{vmatrix} 3 & -1 & -1 \\ -1 & -1 & -1 \\ -1 & 0 & 2 \end{vmatrix} = 8.$$

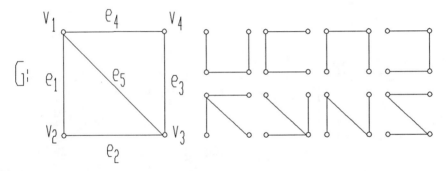

Figure 6.4 $K_4 - e$ and its spanning trees.

The number of labeled trees with p nodes is easily found by applying the Matrix-Tree Theorem to K_p. Each principal cofactor is the determinant of order $p - 1$:

$$\begin{vmatrix} p\text{-}1 & -1 & \cdots & -1 \\ -1 & p\text{-}1 & \cdots & -1 \\ \vdots & \vdots & \ddots & \vdots \\ -1 & -1 & \cdots & p\text{-}1 \end{vmatrix}.$$

Subtracting the first row from each of the others and then adding the last $p - 2$ columns to the first yields an upper triangular matrix whose determinant is p^{p-2}.

Corollary 6.6   The number of labeled trees with p nodes is p^{p-2}. □

There appear to be as many ways of proving this formula as there are independent discoveries thereof. A fascinating compilation of such proofs is presented in Moon [M10].

Cycle Matrix

Let G be a graph whose edges and cycles are labeled. The *cycle matrix* $C = [c_{ij}]$ of G has a row for each cycle and a column for each edge with

$$c_{ij} = \begin{cases} 1 & \text{if the ith cycle contains edge x_j} \\ 0 & \text{otherwise.} \end{cases}$$

In contrast to the adjacency matrix and incidence matrices, the cycle matrix does not determine a graph up to isomorphism. Obviously, the presence or absence of edges which lie on no cycle is not indicated. Even when such edges are excluded, however, C does not determine G as shown

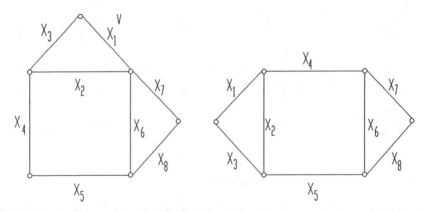

Figure 6.5   Two graphs with the same cycle matrix.

by the pair of graphs in Figure 6.5 which both have cycles

$$Z_1 = \{e_1, e_2, e_3\} \qquad Z_2 = \{e_2, e_4, e_5, e_6\}$$
$$Z_3 = \{e_6, e_7, e_8\} \qquad Z_4 = \{e_1, e_3, e_4, e_5, e_6\}$$
$$Z_5 = \{e_2, e_4, e_5, e_7, e_8\} \quad Z_6 = \{e_1, e_3, e_4, e_5, e_7, e_8\}$$

and therefore share the same cycle matrix

$$
C =
\begin{array}{c}
 \\ z_1 \\ z_2 \\ z_3 \\ z_4 \\ z_5 \\ z_6
\end{array}
\begin{array}{c}
\begin{array}{cccccccc}
e_1 & e_2 & e_3 & e_4 & e_5 & e_6 & e_7 & e_8
\end{array} \\
\left(
\begin{array}{cccccccc}
1 & 1 & 1 & 0 & 0 & 0 & 0 & 0 \\
0 & 1 & 0 & 1 & 1 & 1 & 0 & 0 \\
0 & 0 & 0 & 0 & 0 & 1 & 1 & 1 \\
1 & 0 & 1 & 1 & 1 & 1 & 0 & 0 \\
0 & 1 & 0 & 1 & 1 & 0 & 1 & 1 \\
1 & 0 & 1 & 1 & 1 & 0 & 1 & 1
\end{array}
\right)
\end{array}
$$

The next theorem provides a relationship between the cycle and incidence matrices. In combinatorial topology, this result is described by saying that the boundary of the boundary of any chain is zero.

Theorem 6.7   If G has incidence matrix B and cycle matrix C, then

$$CB^t \equiv 0 \pmod 2.$$

Proof   Consider the ith row of C and the jth column of B^t, which is the jth row of B. The rth entries in these two rows are both nonzero if and only if e_r is in the ith cycle Z_i and is incident with v_j. If e_r is in Z_i, then

v_j is also, but if v_j is in the cycle, then there are two edges of Z_i incident with v_j so that the (i,j)-entry of CB^t is $1 + 1 \equiv 0 \pmod 2$.   □

When we discuss graphs algorithms in Chapter 11, we shall describe data structures related to the adjacency and incidence matrices which are used to save computer memory and implement distance related algorithms more effectively.

EXERCISES 6.2

1. If G is a disconnected labeled graph, then every cofactor of M is 0.   (Brooks, Smith, Stone, and Tutte [BSST1])

2. If G is connected, the number of spanning trees of G is the product of the number of spanning trees of the blocks of G.
    (Brooks, Smith, Stone, and Tutte [BSST1])

3. Do there exist two graphs G_1 and G_2 with the same cycle matrix which are smaller than those in Figure 6.5?

4. For a connected graph G, the rank of the incidence matrix B is $p-1$.

5. For a connected graph G, the rank of the cycle matrix C is $q-p+1$.

6.3   THE DISTANCE MATRIX

Let G be a connected graph whose nodes are labeled $v_1, v_2, \ldots, v_p$. The *distance matrix* $D(G) = [d_{ij}]$ of G has $d_{ij} = d(v_i, v_j)$, the distance between v_i and v_j. Thus $D(G)$ is a symmetric $p \times p$ matrix. Unlike the adjacency, incidence, and cycle matrices, the distance matrix is not a binary matrix. A graph and its distance matrix are shown in Figure 6.6.

Research on distance matrices has been concentrated in two areas: realizability questions and properties of the characteristic polynomial of $D(G)$.

Distance Matrix Realizability

For a matrix D of real numbers to be the distance matrix of a graph G, it must certainly be symmetric with zeros in the diagonal and satisfy the triangle inequality:
$$d_{ij} \leq d_{ik} + d_{kj}.$$

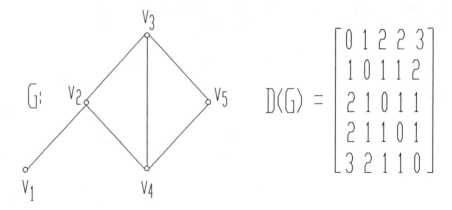

Figure 6.6   A labeled graph and its distance matrix.

Distance matrices are generally considered in the context of weighted graphs. A *weighted graph* consists of a graph together with a weight w_i assigned to each edge e_i. In particular, every graph can be regarded as a weighted graph in which the edges to each have weight one. To see why weighted graphs arise here (besides their usefulness in the related applications), considered the matrix:

$$D = \begin{pmatrix} 0 & 2 & 1 & 1 & 2 \\ 2 & 0 & 3 & 1 & 2 \\ 1 & 3 & 0 & 2 & 1 \\ 1 & 1 & 2 & 0 & 1 \\ 2 & 2 & 1 & 1 & 0 \end{pmatrix}$$

Is D the distance matrix of a graph? As we are restricted to graphs, the answer is easy! Since each edge in G must have weight one, each unit entry in D corresponds to an edge in G. Thus the adjacencies in G are determined by the ones in D and nonadjacencies are given by the other entries of D. Hence we immediately obtain the adjacency matrix A. Since there is a unique labeled graph G corresponding to A, one need only verify whether G is connected. It is easy to check that D is the distance matrix of the sequential join $K_1 + K_1 + \overline{K}_2 + K_1$, appropriately labeled.

Aside from weights on the edges, a distance matrix realization differs from what one usually considers a realization to be. In general one is not looking for a graph G with distance matrix D, but a supergraph H of G such that for any two nodes $u, v \in G$, their distance $d(u, v)$ in H is d_{ij}. Nodes in G are called *main nodes* while those in $H - G$ are called *auxiliary nodes*. The weight $W(H)$ of such a realization H of D is defined

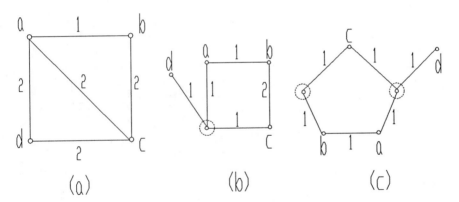

Figure 6.7   Three realizations of D^*.

as the sum of the weights of the edges of H.

$$W(H) = \sum_H w_i.$$

A realization is *optimal* if $W(H)$ is minimum among all possible realizations H of D.

To get a better understanding of these concepts, we illustrate them with the three realizations in Figure 6.7 for the matrix

$$D^* = \begin{pmatrix} 0 & 1 & 2 & 2 \\ 1 & 0 & 2 & 3 \\ 2 & 2 & 0 & 2 \\ 2 & 3 & 2 & 0 \end{pmatrix}$$

The graphs A, B, and C in Figure 6.7 are realizations using 0, 1, and 2 auxiliary nodes, respectively. Their weights are $W(A) = 9$, $W(B) = 6$, and $W(C) = 6$. Note that C is simply graph B with edge bc bisected at a new auxiliary node. Clearly, for any realizable positive integral distance matrix, there is a realization which uses units only. Graph B is preferred since it uses fewer auxiliary nodes.

After several attempts, the reader may find only the three realizations in Figure 6.7 for matrix D^*. One reason for this is a lifelong preference for integers. In fact, neither B nor C is an optimal realization for D^*. In Figure 6.8, we show a realization H^* with weight $W(H^*) = 5$ which is optimal for D^*, in which not all the weights are integers.

We shall say why we know that H^* is an optimal realization presently. Early results on optimal realizations of distance matrices include the following two results of Hakimi and Yau [HY1].

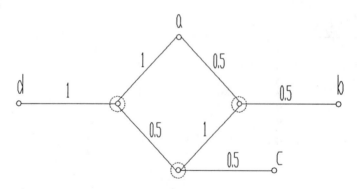

Figure 6.8   An optimal realization for D^*.

Theorem 6.8   No optimal realization contains a triangle.   □

Theorem 6.9   If D has a realization which is a tree, then this realization is optimal.   □

Much of the work on realizations of distance matrices has focused on *tree-realizations*, that is, realizations which are trees. Simões-Periera [S9] characterized those distance matrices which have a tree realization.

Theorem 6.10   Matrix D has a tree-realization if and only if its principal submatrices of order 4 have tree-realizations.   □

Thus, Simões-Periera reduced the problem to determining whether 4×4 matrices have tree realizations. This question was settled by the following result obtained independently by Buneman [B29] and Imrich [I1].

Theorem 6.11   A matrix of order 4 is tree-realizable if and only if among the three sums $d_{12} + d_{34}$, $d_{13} + d_{24}$, and $d_{14} + d_{23}$, two are equal and not smaller than the third.   □

Using this theorem, we can verify that matrix

$$D^* = \begin{pmatrix} 0 & 1 & 2 & 2 \\ 1 & 0 & 2 & 3 \\ 2 & 2 & 0 & 2 \\ 2 & 3 & 2 & 0 \end{pmatrix}$$

used earlier has no tree-realization because none of the sums $d_{12}+d_{34} = 3$, $d_{13} + d_{24} = 5$, and $d_{14} + d_{23} = 4$ are equal. However, Simões-Periera and Zamfirescu [SZ1] found the optimal realization of a non-tree-realizable distance matrix.

Theorem 6.12   The optimal realization of a non-tree-realizable distance matrix with main nodes v_1, v_2, v_3, v_4 consists of a cycle u_1, u_2, u_3, u_4, u_1 of auxiliary nodes with opposite sides having equal weights, plus edges $u_i v_i$ with weight $t_i \geq 0$.   □

If $t_i = 0$ for a given i in Theorem 6.12, then $u_i = v_i$. Thus, the optimal realization uses four or fewer auxiliary nodes. Theorem 6.12 implies that graph H^* in Figure 6.8 is optimal for D^*. Theorems 6.10 and 6.11 provide a method to determine whether a given distance matrix has a tree realization. If it does and a tree realization T is found, then Theorem 6.9 guarantees that T is optimal. If the matrix has no tree realization, the situation is more difficult. However, one can test any realization which is found for optimality using a rather complicated criterion of Simões-Periera [S10].

The Characteristic Polynomial of D

In this subsection, we briefly describe several results which relate to the characteristic polynomial of the distance matrix of G,

$$\Psi_D(x) = \det(D(G) - xI) = \sum_{k=0}^{p} b_k x^k.$$

For the characteristic polynomial

$$\Psi_A(x) = \det(A(G) - xI) = \sum_{k=0}^{p} a_k x^k$$

of the adjacency matrix, Moshowitz [M13] found that the coefficients a_k in $\Psi_A(x)$ for a tree T are given by

$$a_{p-2t} = (-1)^{t+p} M(t) \quad \text{and}$$

$$a_{p-2t-1} = 0,$$

where $M(t)$ denotes the number of ways of selecting t disjoint edges from T. With this motivation, Edelberg, Garey, and Graham [EGG1] investigated the coefficients b_k to see if a similar relationship holds for distance matrices. Earlier, Graham and Pollack [GP1] had determined the constant term in the polynomial $\Psi_D(x)$.

Theorem 6.13   For a tree T on p nodes, the constant in the characteristic polynomial of the distance matrix is

$$b_0 = \det(D(T)) = (-1)^{(p-1)}(p-1)2^{(p-2)}.$$ $\square$

In [EGG1], Edelberg, Garey, and Graham extended this by showing that the terms b_k, for $1 \le k \le 4$ can be found by counting subtrees of various types. The *coup de grace* on this question was obtained by Graham and Lovász [GL2] by extending the work in [EGG1] to show that all the coefficients can be found by counting subforests of various types.

Theorem 6.14   Let F be a forest, $\{T_i\}$ be the set of trees composing F, $p[F]$ and $q[F]$ be its number of nodes and edges, respectively, and let $n_F(T)$ be the number of occurrences of tree T in F. Then for a tree T on p nodes, the coefficient b_k in the expansion of the characteristic polynomial of the distance matrix $D(T)$ is given by

$$b_k = (-1)^{(p-1)}2^{(p-k-2)}\sum_F c_k n_F(T),$$

where the sum ranges over subforests F of T with $k-1$, k, or $k+1$ nodes, and the integer coefficients $c_F(k)$ are

$$c_k = \begin{cases} 4\pi(F)\left(\sum_{i=1}^t \frac{q[T_i]}{p[T_i]} - 1\right) & q[F] = k-1 \\ 4\pi(F)\left(q[F] - \sum_{i=1}^t \frac{1}{p[T_i]}\sum_{x,y\in T_i} d(x,y)\right) & q[F] = k \\ \pi(F)\left(\frac{1}{2}q[F] - \sum_{i=1}^t \frac{1}{[T_i]}\sum_{x,y\in T_i}(d(x,y) - 1)^2\right) & q[F] = k+1 \end{cases}\square$$

By generating $\Psi_D(T)$ for trees T it appears that one always gets distinct polynomials. This led Edelberg, Garey, and Graham [EGG1] to conjecture that if $T \not\cong T'$, then $\Psi_D(T) \not\cong \Psi_D(T')$. McKay [M4] disproved this conjecture by showing that the proportion of trees T, whose distance matrix does not have a cospectral mate goes to zero as $p \to \infty$. Earlier, Schwenk [S5] had discovered the corresponding result for the characteristic polynomial $\Psi_A(T)$ of the adjacency matrix, thereby showing that almost all trees have a cospectral mate.

In this chapter, we have discussed several matrices and algebraic results relating to graphs, focusing on distance concepts. We remark that we have only scratched the surface on this subject. In fact, a whole book could be (and was, see Biggs [B8]) written on algebraic graph theory.

EXERCISES 6.3

1. Find the distance matrix of W_n.

2. Let $D_i(a)$ be the matrix obtained from D by subtracting the non-negative number a from each entry except d_{ii} in the ith row of D. Matrix $D_i(a)$ is tree-realizable if and only if D is.

                                (Simões-Periera and Zamfirescu [SZ1])

3. Find an optimal realization of the matrix

$$D = \begin{pmatrix} 0 & 1 & 2 & 3 \\ 1 & 0 & 2 & 3 \\ 2 & 2 & 0 & 1 \\ 3 & 3 & 1 & 0 \end{pmatrix}$$

4. Determine the characteristic polynomial of the distance matrix of the tree $K_{1,3}$.

FURTHER RESULTS

1. Graham, Hoffman, and Hosoya [GHH1] showed how the determinant of the distance matrix of a directed graph D can be explicitly expressed in terms of the determinants of the blocks of D. We present their result in Chapter 10.

2. Let $c_n(G)$ is the number of n-cycles of a graph G with adjacency matrix A. Then

   a. $c_4(G) = (1/8)[tr(A^4) - 2q - 2\sum_{i \neq j} a_{ij}^{(2)}]$.

   b. $c_5(G) = (1/10)[tr(A^5) - 5tr(A^3) - 5\sum_{i=1}^{p} a_{ii}^{(3)}(\sum_{j=1}^{p} a_{ij} - 2)]$.

                                (Harary and Manvel [HM2])

Suppose that A is the adjacency matrix of G with eigenvalues $\lambda_1, \lambda_2, \ldots, \lambda_t$, where $|\lambda_i| \geq |\lambda_{i+1}|$. If G is k-regular, then $\lambda_1 = k$.

Recently, Chung [C12] showed that if $|\lambda_2|$ is relatively small compared to k, then the diameter of G is also small.

Farrell [F2] developed certain polynomials relating to the cycles in a graph. Among other things, he showed how one might simplify the calculations of the characteristic polynomial using the cycle polynomial which is easier to generate.

Simões-Periera [S12] has just published an algorithm to help obtain optimal realizations for non-tree realizable distance matrices.

CHAPTER 7

Convexity

As with most other chapters, a discussion of convexity could fill a whole book. Indeed, this has been true for many decades, and much has been discovered more recently. We begin by discussing convexity and numerous new graphical invariants it inspired. Next we consider metrics used for graphs besides the usual distance metric. We then focus our attention on various convexity concepts related to geodesics in graphs. Finally, we discuss the concept of distance heredity and associated properties for graphs.

7.1 CLOSURE INVARIANTS

Over the years, various distance-related closure properties have been developed. In most instances, the closure operations involve finding the convex hull of some specified set of nodes under the given metric. Sometimes the operation is iterated until stability occurs. In this section, we discuss several distance-related closure operations on graphs.

The Geodetic Iteration Number

The *closure* (S) of a set S of nodes consists of the nodes of S together with all nodes on geodesics between two nodes of S. Set S is *convex* if all nodes on any geodesic between two of its nodes are contained in S. Thus S is convex if $(S) = S$. The process of taking closures can be repeated to obtain a sequence $S^1, S^2, \ldots$ of geodetic closures, where

$$S^1 = (S), \quad S^2 = ((S)) = (S^1),$$

and in general $S^k = (S^{k-1})$. Since $V(G)$ is finite, the process must terminate with some smallest n for which $S^n = S^{n-1}$. Call the resulting set $[\![S]\!]$, the *convex hull* of S; it corresponds to the smallest convex set containing S, and the value of n is the *geodetic iteration number*, $gin(S)$. For a graph G, $gin(G)$ is now defined as the maximum value of $gin(S)$ over all $S \subset V(G)$.

Determining $gin(G)$ is rather tricky. To check one's understanding of this concept, it is instructive to verify that $gin(K_{2,3}) = 2$. The geodetic iteration number was studied by Harary and Nieminen [HN2] who determined the minimum order of a graph G such that $gin(G) = n$.

Theorem 7.1 Let H_n be any graph having the minimum number of nodes with $gin(H_n) = n$. Then the number of nodes in H_n is 1 if $n = 0$, 3 if $n = 1$, and $n + 3$ if $n \geq 2$. $\square$

The exact structure of these graphs is also determined in [HN2]. Graphs $H_0 = K_1$, $H_1 = K_3$, and $H_2 = K_{2,3}$. The general case is illustrated in Figure 7.1, where v_0^* is adjacent to each odd labeled node, v_1 is adjacent to each even labeled node, $v_2 v_1^*, v_2 v_3 \in E(H_n)$, and $v_{2k} v_{2k-1}, v_{2k} v_{2k+1} \in E(H_n)$.

Convex hulls were also used by Nieminen to characterize trees and complete graphs [N4].

The Geodetic Number

A *geodetic cover* of G is a set $S \subset V(G)$ such that every node of G is contained in a geodesic joining some pair of nodes in S. The *geodetic number* $gn(G)$ of G is the minimum order of its geodetic covers, and any cover of order $gn(G)$ is a *geodetic basis*. For a geodetic cover S, we call a node of $G - S$ a *check node*. Some simple observations about $gn(G)$ for nontrivial graphs are that

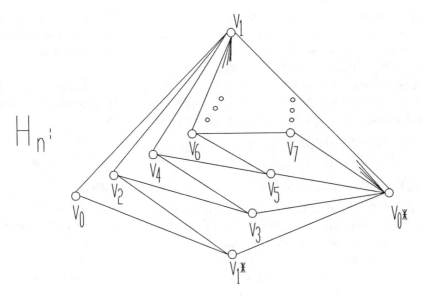

Figure 7.1   Minimum order graph H_n with $gin(H_n) = n$.

1. $2 \leq gn(G) \leq p$ and $gn(K_p) = p$.
2.   If $\deg v \leq 1$, then v is in every geodetic basis of G.
3.   Every cutnode is a check node of every geodetic basis.

Buckley, Harary, and Quintas [BHQ1] characterized those connected graphs G for which $gn(G) = p$, $p - 1$, or 2.

Theorem 7.2   Let G be a connected graph. Then

1. $gn(G) = p$ if and only if $G = K_p$.
2. $gn(G) = p - 1$ if and only if $G = K_1 + \cup m_j K_j$, where $2 \leq \sum m_j$. $\square$

The graphs with $gn(G) = p - 1$ consist of the join of two or more complete graphs with K_1 and the cutnode is the only check node. A *diametral path* is a geodesic of length $d(G)$ joining peripheral nodes u and v. Hence, a diametral path is a path of length $d(G)$ that is an eccentric path for its endnodes.

Theorem 7.3   For a connected graph G, $gn(G) = 2$ if and only if there exist peripheral nodes u and v such that every node of G is on a diametral path joining u and v.

Proof   Let u and v be nodes such that each node of G is on a diametral path P joining u and v. Since G is nontrivial, $gn(G) \geq 2$. Since P is a geodesic joining u and v, each node of G is on a geodesic between u and v, so $S = \{u, v\}$ is a basis and $gn(G) = 2$.

Conversely, let $gn(G) = 2$ and $S = \{u, v\}$ be a basis for G. If $d(u, v) < d(G)$ then there exist nodes s and t on distinct geodesics joining u and v such that $d(s, t) = d(G)$. But then we get

(1) $d(u, v) = d(u, s) + d(s, v)$,

(2) $d(u, v) = d(u, t) + d(t, v)$, but

(3) $d(s, t) \leq d(s, u) + d(u, t)$ by the triangle inequality.

Since $d(u, v) < d(G) = d(s, t)$, (3) implies $d(s, t) < d(s, u) + d(u, t)$, which together with (1) yields $d(s, v) < d(u, t)$. Then using (3) with v playing the role of u gives, by (2),
$d(s, t) \leq d(s, v) + d(v, t) < d(u, t) + d(v, t) = d(u, t) + d(t, v) = d(u, v)$.
Thus $d(s, t) < d(u, v)$, a contradiction. Hence $d(u, v) = d(G)$ and each node of G is on a diametral path joining u and v.   □

Obviously, the unique geodetic basis of a tree consists of all its end-nodes. In [BHQ1], the value of $gn(G)$ was determined for various classes of graphs such as unicyclic graphs, complete multipartite graphs, and prisms of an n-cycle. They also solved various extremal problems involving $gn(G)$ by determining the maximum and minimum values for $gn(G)$ among graphs having p nodes or q edges, as well as the minimum when both p and q are given. Since almost all graphs have diameter two, the following result often proves useful.

Theorem 7.4   If $d(G) = 2$ and G contains an independent set I of nodes such that each node in $V(G) - I$ has at least two neighbors in I, then $gn(G) \leq |I|$.   □

For example, using Theorem 7.4 it becomes easy to show that the Petersen graph has geodetic number 4.

Two classes of graphical games called *achievement and avoidance games* were presented by Harary in [H16]. These games were examined for the geodetic number by Buckley and Harary [BH3] and Necásková [N3].

Suppose that instead of looking for a basis, we select nodes sequentially as follows. Select a node v_1 and let $S_1 = \{v_1\}$. Select $v_2 \neq v_1$, let $S_2 = \{v_1, v_2\}$ and then successively select node $v_k \notin (S_{k-1})$. The *closed geodetic number* $cgn(G)$ is the smallest k for which selection of v_k in the given manner makes $(S_k) = V(G)$. Games related to this invariant were studied in [BH2].

Another related invariant is the sequential geodetic number defined by Harary [H16]. The process begins as above but differs after S_2. Select a node v_1 and let $S_1 = \{v_1\}$. Select $v_2 \neq v_1$, let $S_2 = \{v_1, v_2\}$ and then successively select $v_k \notin S_{k-1}$ and let

$$S_k = S_{k-1} \cup \{v_k\} \cup \{\text{nodes on } v_k\text{–}u \text{ geodesics for } u \in S_{k-1}\}.$$

The *sequential geodetic number* $sgn(G)$ is the smallest k such that there is a sequence $(v_1, v_2, \ldots, v_k)$ for which $S_k = V(G)$.

At first it may seem that $cgn(G) = sgn(G)$. For $sgn(G)$ the sets S_k only include new nodes v_k and nodes on geodesics of length $d(v_k, S_{k-1})$; whereas, for $cgn(G)$ nodes on geodesics of all lengths between v_k and nodes in S_{k-1} are included in S_k. To check one's understanding of the difference, it is instructive to verify that for the wheel $W_{1,6}$, $cgn(W_{1,6}) = 3$ but $sgn(W_{1,6}) = 4$.

The Hull Number

The *hull number* $h(G)$ is the minimum order of a set $S \subset V(G)$ such that its convex hull $[\![S]\!]$ is $V(G)$. Such a set is called a *minimum hull set*. Since the convex hull can be found by repeatedly taking closures until stability occurs, it is clear that $h(G) \leq gn(G)$. This observation and Theorem 7.2 lead easily to the conclusion that $h(G) = p$ or $p - 1$ just for the graphs K_p and $K_1 + \cup m_j K_j$, where $2 \leq \sum m_j$.

For a convex set $S \subset V(G)$, node $v \in S$ is an *extreme node* of S if $S - v$ is also convex. The hull number was studied by Everett and Seidman [ES3] who obtained a number of bounds for $h(G)$. The following two are typical of the simpler results.

Theorem 7.5 If G is a connected graph of order p with k extreme nodes, then $k \leq h \leq p - d + 1$.

Proof The lower bound follows from the fact that each extreme node must be in each minimum hull set. Since the $d - 1$ internal nodes of any diametral path can be eliminated from S simply by including the endnodes of that path, the stated upper bound holds.   □

Theorem 7.6 If G is n-connected, then $h \leq p - n\lfloor d/2 \rfloor$.   □

Everett and Seidman obtained other bounds relating to extreme nodes and neighborhood sets of cliques. They observed that the minimum hull set consisting entirely of extreme nodes is unique. However, even when

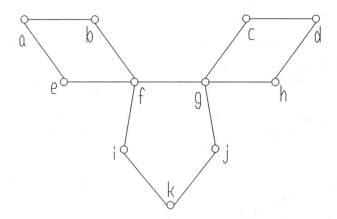

Figure 7.2   A graph with no extreme nodes in its minimum hull set.

there are no extreme nodes in a minimum hull set, it may still be unique as illustrated in Figure 7.2 where none of the nodes a, d, or k is extreme, but $\{a, d, k\}$ is the unique minimum hull set.

Other Geodetic Closure Invariants

There are other invariants related to geodetic closures which have been investigated in varying degrees. In fact, any type of abstract convexity can generally be translated into at least one and often several interesting graphical invariants to study. The *caratheodory number* $cara(G)$ of a graph G is the smallest positive integer k such that for all sets $S \subset V(G)$ and $v \in [\![S]\!]$, there exists $S' \subset S$ with $|S'| \leq k$ such that $v \in [\![S']\!]$.

A set S of nodes is *monophonically-convex* if S contains every node on every induced path between nodes in S. A slight variant of the caratheodory number was studied for monophonic-convexity by Farber and Jamison [FJ1].

A set $S \subset V(G)$ has a *radon partition* $S = S_1 \cup S_2$ if $[\![S_1]\!] \cap [\![S_2]\!] \neq \emptyset$, that is, if the intersection of their convex hulls is nonempty. The *radon number* $radon(G)$ of G is the least integer k such that every subset $S \subset V(G)$ with $|S| \geq k$ has a radon partition. Since K_p has no radon partition, the definition is extended to all graphs by defining $radon(K_p) = p+1$. The radon number was studied by Delire [D2] who classified all graphs G with $p - 2 \leq radon(G) \leq p + 1$.

The *helly number* $hell(G)$ of a graph G is the smallest integer k such that for any family $\mathfrak{I}$ of convex sets of nodes from G, all its subfamilies

$\mathfrak{S}'$ of convex sets of order k satisfy

$$\bigcap_{S_i' \in \mathfrak{S}'} S_i' \neq \emptyset \quad \text{implies} \quad \bigcap_{S_i \in \mathfrak{S}} S_i \neq \emptyset.$$

By studying the radon number, one obtains corresponding information about the helly number according a theorem of Kay and Womble [KW1] which asserts that $radon(G) = 1 + hell(G)$.

EXERCISES 7.1

1. Determine $gn(G)$ for each of the following:
   a. C_p b. Q_n c. $K_{m,n}$
2. Determine $cgn(G)$ for each of the following:
   a. C_p b. Q_n c. $K_{m,n}$
3. Determine $sgn(G)$ for each of the following:
   a. C_p b. Q_n c. $K_{m,n}$
4. The following relations hold: $gn(G) \leq cgn(G) \leq sgn(G)$.
5. Determine $gin(G)$ for each of the following:
   a. C_p b. Q_n c. $K_{m,n}$
6. Construct a smallest order graph G for which $h(G) < gn(G)$.
7. Determine $h(G)$ for each of the following:
   a. C_p b. Q_n c. $K_{m,n}$
8. Construct a graph with $gin(G) = 4$ and $gn(G) = 2$.

7.2   METRICS ON GRAPHS

The standard metric used on graphs is the distance metric, $d(u,v)$, the length of a shortest path joining nodes u and v, which may be called the *path metric*. For this metric, many interesting questions arise as we have seen throughout the text. In Chapter 6, we discussed the problem of realizing a given distance matrix by a graph. We now consider a related question — embedding problems for graphs.

Isometric Embeddings

For two connected graphs G_1 and G_2 with distance metrics d_1 and d_2, respectively, a set of nodes $S_1 \subset V_1$ is *isometrically embeddable* in G_2 if there is a set $S_2 \subset V_2$ and a bijection $f : S_1 \longrightarrow S_2$ such that for all $u, v \in S_1$, $d_1(u, v) = d_2(f(u), f(v))$. This concept was examined for trees and bipartite graphs by Melter and Tomescu [MT1]. They used the following two lemmas to characterize trees in terms of isometric embeddings.

Lemma 7.7a   For any three nodes u, v, and w of a tree
$d(u, v) + d(v, w) + d(x, w) \equiv 0 \pmod 2$.   □

Lemma 7.7b   For any four nodes u, v, w, and x in a tree, the numbers $d(u, v) + d(w, x)$, $d(u, w) + d(v, x)$,   and $d(u, x) + d(v, w)$ are not all distinct.   □

Theorem 7.7   A connected graph G is a tree if and only if every subset $G_S \subset V$ of at at most 4 nodes is isometrically embeddable in some tree.

Proof   The necessity is trivial so we need prove only the sufficiency.

Assume that G contains an odd cycle and let C be the shortest with $2k+1$ nodes. Let v and w be adjacent nodes on C, and let u be the unique node on C such that $d(u, v) = d(u, w)$. Then

$$d(u, v) + d(v, w) + d(u, w) = 2k + 1 \equiv 1 \pmod 2.$$

But then Lemma 7.7a implies that these three nodes cannot isometrically embed in a tree, a contradiction, so G must be bipartite.

Next assume that G contains an even cycle C' of length $2k$ ($k > 2$). Let u and x be antipodes on C', and let v and w be antipodal nodes which are neighbors of u and x, respectively. Then $d(u, v) + d(w, x) = 2$, $d(u, w) + d(v, x) = 2k - 2$, and $d(u, x) + d(v, w) = 2k$, which are all distinct. Lemma 7.7b then asserts that these four nodes cannot isometrically embed in a tree.

The only remaining case is $G = C_4$. But clearly the nodes of C_4 do not isometrically embed in a tree. Hence G is connected and has no cycles, so G is a tree.   □

It is interesting to compare Theorems 7.7 and 6.10 and note that Simões-Pereira's characterization of tree realizable distance matrices depends only on the principal submatrices of order 4.

Winkler [W5] surveys results concerning isometric embeddings of graphs in cartesian products. The hypercube Q_n is the cartesian product of $n - 1$ copies of K_2 and as noted in [W5] is of interest in relation to addressing schemes for communications networks. Let N_{uv} denote the set of nodes nearer to x than to y in G. Djokovič [D10] characterized graphs isometrically embeddable in Q_n.

Theorem 7.8   A connected graph G can be isometrically embedded in Q_n if and only if G is bipartite and for each edge $uv \in G$, both N_{uv} and N_{vu} are convex. $\square$

Euclidean Embeddings

A *euclidean embedding* of a graph G is a mapping of G into $\Re^n$ such that adjacent nodes have euclidean distance 1 and nonadjacent nodes have some other distance. Erdös, Harary and Tutte [EHT1] defined the *euclidean dimension* $ed(G)$ of a graph G to be the smallest n so that G has a euclidean embedding into $\Re^n$. It is easy to verify that $ed(K_p) = p - 1$. Figure 7.3 shows euclidean embeddings of the Petersen graph P into $\Re^3$ and $\Re^2$. Since there is no euclidean embedding of P into $\Re^1$, $ed(P) = 2$.

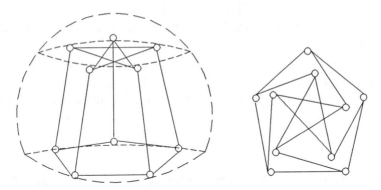

Figure 7.3   Two euclidean embeddings of the Petersen graph.

The invariant $ed(G)$ is nondecreasing, that is, if H is an induced subgraph of G, then $ed(H) \leq ed(G)$. However, a homeomorph of a graph may have a smaller value of ed. So for example, although $ed(K_5) = 4$ and $ed(K_{3,3}) = 4$, there are nonplanar graphs G with $ed(G) = 2$.

Buckley and Harary [BH4] determined the euclidean dimension of wheels, generalized wheels, and complete tripartite graphs. The following result of Maehara [M2] subsumed the last of these results by determining a formula for the euclidean dimension of any complete multipartite graph.

Theorem 7.9   Let G be a complete multipartite graph with s parts of size 1, t parts of size 2, and u parts of size greater than 2. Then

$$ed(G) = \begin{cases} s + t + 2u & \text{if } t + u \geq 2; \\ s + t + 2u - 1 & \text{if } t + u \leq 1. \end{cases} \qquad \square$$

The Metric Dimension

A *metric basis* of a connected graph G is a smallest set $S = \{v_1, v_2, \ldots, v_n\}$ of nodes of G such that for nodes $u \in G$, the ordered n-tuples of distances $[d(u, v_1), d(u, v_2), \ldots, d(u, v_n)]$ are all distinct. The *metric dimension* of G is the cardinality of a metric basis for G. For trees T, Slater and independently Harary and Melter [HM3] found an algorithm to determine a metric basis for T, gave an explicit formula for the metric dimension of T, and showed that every tree has a metric basis consisting of endnodes only.

In an analogous study, Melter and Tomescu [MT2] examined the boolean metric dimension of a graph where the *boolean distance* $bd(x, y)$ is $\emptyset$ if $x = y$ and $\{z : z$ lies on some path between x and $y\}$ if $x \neq y$. Note that the boolean distance is a *set*, not its cardinality. In this situation, the conditions of a metric are described in terms of equality, containment, and unions of sets.

By taking the multiplicity of boolean distances into account, we obtain the boolean distance multiset $B(G)$ of G, which lists each boolean distance as well as how many times it occurs for some pair of nodes of G. Harary, Melter, Peled, and Tomescu [HMPT1] determined when a graph can be reconstructed by using $B(G)$

Theorem 7.10   A nontrivial connected graph G is reconstructible from $B(G)$ if and only if every block of G is either an edge or a triangle. $\square$

Other Metrics

Besides the usual path metric that describes the length of a shortest path joining a pair of nodes, other metrics have been used to indicate the distance between sets of nodes in a graph or the distance between graphs themselves. For example, Chartrand, Oellermann, Tian and Zou [COTZ1] define the *Steiner distance* $sd(S)$ of a set S of nodes in a connected graph G as the minimum size of a connected subgraph of

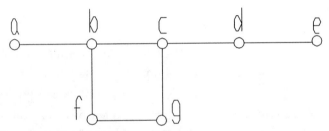

Figure 7.4   A graph to illustrate Steiner distance.

G containing S. For example, consider the graph in Figure 7.4. Let $S_1 = \{d, f, g\}$ and $S_2 = \{a, e, f\}$. Then $sd(S_1) = 3$ and $sd(S_2) = 5$.

Note that if H is a connected subgraph of G for which $S \subset V(H)$ and $q(H) = sd(G)$, then H is a tree, often called a *Steiner tree*. In [COTZ1] the *n-eccentricity* $e_n(v)$ of a node is defined as the maximum Steiner distance among n-element sets containing v. Then the *n-radius* $r_n(G)$ and *n-diameter* $d_n(G)$ are defined naturally. Among the results in [COTZ1] is the following.

Theorem 7.11   If T is a tree and $3 \le n \le p$, then $d_{n-1}(T) = r_n(T)$. □

In a subsequent paper, Oellermann and Tian [OT1] defined the *Steiner n-center* $C_n(G)$, $n \ge 2$ as the subgraph induced by the set of nodes v for which $e_n(v) = r_n(G)$. Since the Steiner 2-center $C_2(G)$ is simply $C(G)$, this concept generalizes that of the center. Oellermann and Tian showed that for each $n \ge 2$, every graph is the Steiner n-center of some graph. They also characterized those trees that are Steiner n-centers of trees. Of course, Theorem 2.1 and the fact that $C_2(G) = C(G)$ imply that $C_2(G) = K_1$ or K_2.

Theorem 7.12   Let $n \ge 3$ be an integer and let T be a tree. Then T is the n-center of some tree if and only if T has at most $n - 1$ endnodes. □

One can define *the distance between two graphs*, G and H, both of order p, $\partial(G, H)$, as the minimum order of a graph containing both G and H as induced subgraphs. This concept was studied by Zelinka [Z5].

Theorem 7.13   The function ∂ is a metric on the set of all graphs of order p.   □

EXERCISES 7.2

1. Draw a nonplanar graph G with euclidean dimension $ed(G) = 2$.
2. The euclidean dimension of a wheel $W_{1,n}$ is 2 if $n = 6$ and 3 if $n \neq 6$.
   (Buckley and Harary [BH4])
3. An *equilateral triangle* in a graph is a set of three nodes u, v, and w such that $d(u,v) = d(u,w) = d(v,w)$. A graph G has no equilateral triangles if and only if each component of G is either a path or a cycle whose length is not divisible by three.
   (Harary and Melter [HM4])
4. Determine the metric dimensions of the following graphs:
   a. K_p b. C_p c. $K_{m,n}$ d. $W_{1,n}$
5. Let $md(G)$ denote the metric dimension of G. For connected graphs G_1 and G_2, the metric dimension of the join $G_1 + G_2$ is given by $md(G_1 + G_2) = md(G_1) + md(G_2)$.   (Harary and Melter [HM3])

7.3 GEODETIC GRAPHS

For a tree T, there is a unique path joining any pair of nodes in T. Thus every path in T is a geodesic. A graph G is *geodetic* if each pair of nodes in G is joined by a unique shortest path. Research on geodetic graphs has focused on two main items. Early studies tried to characterize the structure of geodetic graphs of a given diameter, while more recent investigations have dealt with techniques for constructing one geodetic graph from others.

Structural Results

Some early observations of Stemple and Watkins show that it is appropriate to focus on the blocks of a graph to check whether it is geodetic. They showed [SW1] that a graph is geodetic if and only if each of its blocks is geodetic. They also characterized planar geodetic graphs.

Theorem 7.14   A planar graph G is geodetic if and only if each block of G is either K_2, an odd cycle, or a geodetic graph homeomorphic to K_4.
   □

Stemple [S24] gives a structural description of geodetic graphs of diameter two in terms of order, degree, and relationships among the cliques in the graph. Later, Scapellato [S4] built on Stemple's results to show a connection to various geometric structures. He then showed that a 2-connected geodetic graph of diameter 2 is either strongly regular with $\mu = 1$ (see §8.3), a graph called a "pyramid", or a π-graph (the last two graphs are defined in terms of incidence relationships among their cliques).

Plesník [P4] introduced a certain type of graph that links the early characterization work and the later construction processes. He defines a graph $K_n^{(f)}$ in terms of a special type of homeomorph of K_n. Let the nodes of K_n be $v_1, v_2, \ldots, v_n$, and for each i, let $f(v_i) = x_i$ be a nonnegative integer. Then the graph $K_n^{(f)}$ is obtained from K_n by inserting $x_i + x_j$ additional nodes on edge $v_i v_j$ (or equivalently, replacing the edge by a path of length $x_i + x_j + 1$). Plesník showed that these graph are geodetic. His result was extended by Stemple [S25].

Theorem 7.15   A homeomorph of K_n is geodetic if and only if it is of the form $K_n^{(f)}$.   □

As we shall soon see, the graph $K_n^{(f)}$ has been used in a number of constructions of geodetic graphs.

As usual, let $N_i(v) = \{u : d(u,v) = i\}$, that is, the set of nodes in the ith neighborhood of v. Parthasarathy and Srinivasan [PS1] obtained some beautiful constructions of geodetic blocks. In order to check that the resulting graph is indeed geodetic, they developed the following characterization.

Theorem 7.16   A graph G is geodetic if and only if for every node v, each node $u \in N_k(v)$ is adjacent to a unique node in $N_{k-1}(v)$ for $2 \le k \le e(v)$.

Proof   If for an arbitrary pair u, v of nodes in G, each node $u \in N_k(v)$ is adjacent to a unique node in $N_{k-1}(v)$ for $2 \le k \le e(v)$, then by backtracking we obtain a unique geodesic joining u and v. Thus G is geodetic.

On the other hand, assume that G is geodetic and for some node v and some integer k, a node $u \in N_k(v)$ has two neighbors $w_1, w_2 \in N_{k-1}(v)$. Then $d(u,v) = k$ and there are two geodesics joining u and v, one through

w_1 and another through w_2, contradicting the fact that G is geodetic. Hence, each node $u \in N_k(v)$ is adjacent to a unique node in N_{k-1} for $2 \leq k \leq e(v)$.   □

Additional structural results for geodetic blocks are contained in Stemple [S24], Parthasarathy and Srinivasan [PS2] and Alagar and Srinivasan [AS1]. For example, Stemple showed that every geodetic block of diameter 2 is self-centered and each of its nodes lies on an induced C_5. Parthasarathy and Srinivasan showed that every geodetic block of diameter 3 is self-centered and each of its nodes lies on an induced C_7. This is the end of the road as for diameter 4 or greater, a geodetic block need not be self-centered. For example, the graph in Figure 7.5 is a geodetic block of diameter 4 but is not self-centered. Note that this graph corresponds to $K_4^{(f)}$ where $f(v_1) = 0$ and $f(v_k) = 1$ for $k = 2, 3, 4$. Thus by Plesnik's result, Theorem 7.15, the graph is geodetic.

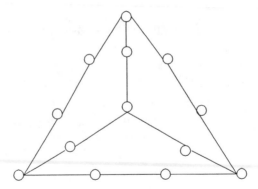

Figure 7.5   A geodetic block that is not self-centered.

Parthasarathy and Srinivasan [PS3] found bounds on the number of edges in geodetic graphs of diameter d and other bounds for geodetic blocks of diameter d that are either self-centered or not. An interesting result in that paper states that in a geodetic block of diameter d which is neither K_2 nor an odd cycle, there are at least 4 nodes of each eccentricity $t, \lfloor \frac{1}{2}(d+2) \rfloor \leq t \leq d$.

Alagar and Srinivasan [AS1] showed that every geodetic block $G \neq C_{2d+1}$ with diameter $d = d(G) \geq 2$ contains an induced $K_{1,3}$. Thus by Theorem 1.17, such a graph cannot be a line graph.

Constructing Geodetic Graphs

Because of Stemple and Watkin's observation that a graph is geodetic if and only if each of its block is geodetic, constructive results in this area have focused mostly on geodetic blocks.

Cook and Pryce [CP1] and Plesník [P6] used a concept called *pulling* to construct a new geodetic graph from a given geodetic graph. The way pulling works is to first split each main node (deg ≥ 3) in a $K_t^{(f)}$ subgraph ($t \geq 3$) for which $f(v_i) \geq 1$ for each main node. Then replace $K_t^{(f)}$ in the geodetic graph by a smaller $K_t^{(f')}$ with $f'(v_i) = f(v_i) - 1$ for each main node. The process can be repeated so long as $f'(v_i) > 0$ for all i. Whenever the resulting graph G^* satisfies some rather simple conditions on the path lengths between triples of its main nodes, the resulting graph G^* is geodetic. See Plesník [P6] for further details on this process and the conditions to check.

Parthasarathy and Srinivasan [PS1] developed a technique for constructing new geodetic graphs from others. Their technique enables the building of geodetic graphs of large diameter. They also modify the construction to produce geodetic homeomorphs of a geodetic block, thus obtaining geodetic blocks with large diameter and girth. Their constuction begins with a clique cover of graph G, that is, a set S of edge-disjoint maximal complete subgraphs of G partitioning $E(G)$. Let the nodes of G be $v_1, v_2, \ldots, v_p$ and label the cliques of S by $S_1, S_2, \ldots, S_t$. Form the graph G^* as follows. For each node v_i of G, let c_i be the number of cliques it is in. Then to build graph G^*, begin by replacing v_i by the star K_{1,c_i}. Now label the center node of K_{1,x_i} by $v_{i,0}$ and label the other nodes by $v_{i,k}$ for the cliques S_k that contain v_i. After labeling the nodes of each star in this manner, insert edges $v_{i,j} v_{i,j'}$ for $j, j' \geq 1$. The resulting graph is G^*. Parthasarathy and Srinivasan showed that if G is a geodetic block, then G^* is also geodetic.

Strongly Geodetic Graphs

If each pair of nodes u, v of a graph G is joined by at most one path of length not more than the diameter $d(G)$, then G is called *strongly geodetic*. Every strongly geodetic graph is geodetic, but not conversely. Strongly geodetic graphs were studied by Bosák, Kotzig, and Znám [BKZ1], who showed that there are two basic types of such graphs.

Theorem 7.17   If G is strongly geodetic, then G is either a forest or a connected regular graph.   □

They also showed that the connected strongly geodetic graphs are precisely the Moore graphs discussed in §5.2.

EXERCISES 7.3

1. If G has no even cycles, then G is geodetic.
2. In a geodetic graph, any shortest cycle must have odd length.
3. If G is a block such that the shortest cycle containing any pair of nodes is odd, then G is geodetic.
4. Draw a geodetic graph that is not strongly geodetic.
5. If for all pairs of nodes $u, v \in G$ such that $d(u, v) \leq p/2$, there is a unique geodesic joining u and v, then G is geodetic.
6. For each odd integer $g \geq 3$ and any integer $d \geq g - 2$, there exists a self-centered geodetic block on $3k + 1$ nodes having girth g and diameter d.   (Parthasarathy and Srinivasan [PS1])

7.4   DISTANCE-HEREDITARY GRAPHS

There are a number of graphs that are defined in terms of convexity and distance relations among pairs, triples, or quadruples of nodes. In this section we discuss several such graph classes.

Convex Basic Graphs

In §7.1, we saw that under the usual distance metric a set S is convex if all the nodes on geodesics joining two nodes of S are also in S. For any graph G, with $V(G) = \{v_1, v_2, \ldots, v_p\}$, the sets $\emptyset$, $\{v_i\}$, and $V(G)$ are always convex. A connected graph G is *convex basic* if those are the only convex sets in G.

The study of convex basic graphs has focused mainly on planar graphs, that is, graphs that can be drawn in the plane with no crossing edges. Planar convex basic graphs were characterized by Hebbare and Rao [HR4].

Theorem 7.18   Let G be a planar graph of order at least 4 other than the cube Q_3. Then G is convex basic if and only if for all paths u, v, w of length two there is a node $z \in G$ such that $\langle \{u, v, w, z\} \rangle$ is C_4.   □

Hebbare and Rao also observed that a planar convex basic graph on at least three nodes must have at least one node of degree two. Later Hebbare [H18] improved this to show that there must in fact be at least two such nodes and described the graphs having exactly two or three nodes of degree two.

Preserving Distance Properties in Spanning Trees

For a connected graph G, there are a number of distance related properties one might wish to preserve in a spanning tree of G. Of course, the distances themselves cannot be preserved unless G is a tree, since once an edge uv is removed, the distance $d(u, v)$ increases. Nandakumar [N1] determined when it is possible to preserve the eccentricities. Let $d = d(G)$ be the diameter and $r = r(G)$ the radius of G.

Theorem 7.19   A connected graph G has an eccentricity-preserving spanning tree if and only if the following conditions hold:

1. $\langle C(G) \rangle = K_1$ and $d = 2r$, or
    $\langle C(G) \rangle = K_2$ and $d = 2r - 1$;   and
2. Each v with $e(v) > r$ has a neighbor u for which $e(u) = e(v) - 1$. □

Rather than requiring all eccentricities to remain unchanged in a spanning tree, one might only ask that the maximum eccentricity not change. A *diameter-preserving spanning tree* of a graph G is a spanning tree T for which $d(T) = d(G)$. Graphs with diameter-preserving spanning trees were characterized elegantly by Buckley and Lewinter [BL1].

Theorem 7.20   A connected graph G has a diameter-preserving spanning tree if and only if either

1. $d(G) = 2r(G)$, or
2. $d(G) = 2r(G) - 1$ and G contains a pair of adjacent center nodes u and v which have no common eccentric node.   □

Another related problem involving spanning trees is to preserve the center of G. A spanning tree T for which $C(T) = C(G)$ is called a *center-preserving spanning tree*. These graphs were discussed in [BL2], where it was noted that no useful nontrivial characterization yet exists for such graphs. Finally, we note that Buckley and Palka [BP3] studied conditions under which a random graph has a diameter-preserving spanning tree.

Distance-Hereditary Graphs

Rather than check whether a graph has a single subgraph (such as a spanning tree) with some distance preserving property, one might ask that all subgraphs in some class share some such property. The graphs we now discuss fall into this category. A graph G is *distance-hereditary* if for all connected induced subgraphs F of G, $d_F(u,v) = d_G(u,v)$ for all pairs of nodes $u, v \in F$. The first characterizations of such graphs were found by Howorka [H26]

Theorem 7.21   The following conditions are equivalent for a graph G:

1. G is distance-hereditary.
2. Every induced path of G is a geodesic.
3. Each cycle C_n, $n \geq 5$, in G has at least two chords and each C_5 in G can be labeled a, b, c, d, e, a so that ac and bd are edges of G. $\square$

Using Theorem 7.21, Buckley and Palka [BP1] and Bandelt and Mulder [BM1] independently obtained the following forbidden induced subgraph characterization of these graphs.

Theorem 7.22   Graph G is distance-hereditary if and only if G contains no C_n, $n \geq 5$, nor any of the graphs in Figure 7.6 as an induced subgraph.

Proof   If G is distance-hereditary, then by Theorem 7.21, G cannot contain an induced C_n, $n \geq 5$. Also, G cannot contain any of the graphs in Figure 7.6 as an induced subgraph since they each contain an induced path that is not a geodesic.

For the converse, suppose G contains no C_n, $n \geq 5$, nor graphs A, B, or C in Figure 7.6 as an induced subgraph. Let C_k be a longest cycle in G having only one chord. If $k \geq 7$, then G contains an induced C_n with $n \geq 5$, a contradiction. Thus $k = 4$, 5, or 6. If $k = 4$ then

$\langle C_k \rangle = K_4 - e$, which is distance-hereditary. If $k = 5$, then G contains an induced subgraph A, a contradiction. If $k = 6$, then G contains either an induced C_5 or graph C, a contradiction. Hence each cycle of length at least 5 must contain at least two chords. Suppose some $C_5 \in G$ has exactly two chords. Then since G does not contain an induced B, it must be possible to label C_5 by a, b, c, d, e, a so that ac and bd are edges of G. Thus by Theorem 7.21, G is distance-hereditary. $\quad\square$

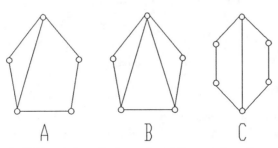

Figure 7.6   Three forbidden induced subgraphs of distance-hereditary graphs.

Theorem 7.22 was used in [BP1] to examine random graphs that are distance-hereditary. Besides discovering Theorem 7.21, Bandelt and Mulder [BM1] obtained several other characterizations. One of their metric characterizations is the following.

Theorem 7.23   A graph G is distance-hereditary if and only if for any four nodes $u, v, w, x \in G$, at least two of the following sums are equal:
$$d(u, v) + d(w, x), \quad d(u, w) + d(v, x), \quad d(u, x) + d(u, w).$$
Furthermore, if it is the smaller distance sums that are equal, then the larger sum exceeds the smaller sum by at most 2. $\quad\square$

Another characterization in [BM1] describes the structure of these graphs.

Theorem 7.24   Let G be a nontrivial connected graph. Then G is distance-hereditary if and only if G can be obtained from K_2 by a sequence of the the following operations:

1. Adding a new node v' and joining it only to one node v.

2. Adding a new node v' and joining it to some node v and all its neighbors.

3. Adding a new node v' and joining it to the neighbors of some node v but not to v. $\quad\square$

Ptolemaic Graphs

A graph G is *ptolemaic* if every quadruple u, v, w, x of its nodes satisfy

$$d(u, v)d(w, x) \leq d(u, w)d(v, x) + d(u, x)d(v, w).$$

These graphs were studied by Howorka [H27] who showed that they are a subset of distance-hereditary graphs.

Theorem 7.25   A graph G is ptolemaic if and only if it is distance-hereditary and contains no induced C_4. $\square$

Using this result, Buckley and Palka [BP1] studied random graphs that are ptolemaic. Bandelt and Mulder [BM1] used Theorem 7.25 and their characterizations of distance-hereditary graphs to obtain a metric characterization as well as a structural characterization of ptolemaic graphs. Relations between chordal graphs (see §2.1) and ptolemaic graphs were obtained by Farber and Jamison [FJ1].

EXERCISES 7.4

1. Draw all convex basic graphs on 5 or fewer nodes.
2. If G is unicyclic, $C(G) = \{u, v\}$, and $uv \in E(G)$, then G has a center-preserving spanning tree if and only if u and v have no common eccentric node.
3. Draw all graphs on six or fewer nodes which are not distance-hereditary.
4. Which wheels are distance-hereditary?
5. Find a smallest distance-hereditary graph that is not ptolemaic.

CHAPTER 8

Symmetry

From its inception, the theory of groups has provided an interesting and powerful abstract approach to the study of the symmetries of various configurations. It is not surprising that there is a particularly fruitful interaction between groups and graphs. In order to place the topic in its proper setting, we recall some elementary but relevant facts about groups. In particular, we develop several operations on permutation groups. A whole book could be written (and has, see White [W2]) concerning the interplay of graphs and groups. We shall focus our attention on those results which relate to the distance concepts we have been considering.

8.1 GROUPS

First we recall the usual definition of a group. The nonempty set $A = \{\alpha, \beta, \gamma, \ldots\}$ together with a binary operation, denoted by juxtaposition, constitutes a *group* whenever the following four axioms are satisfied:

Axiom 1 (closure) For all α, β in A, $\alpha\beta$ is also an element of A

Axiom 2 (associativity) For all α, β, γ in A,

$$\alpha(\beta\gamma) = (\alpha\beta)\gamma.$$

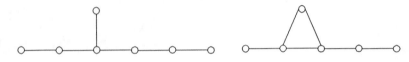

Figure 8.1   Two identity graphs.

Axiom 3 (identity) There is an element ι in A such that

$$\iota\alpha = \alpha\iota = \alpha \qquad \text{for all } \alpha \text{ in } A.$$

Axiom 4 (inversion) If axiom 3 holds, then for each α in A, there is an element denoted α^{-1} such that

$$\alpha\alpha^{-1} = \alpha^{-1}\alpha = \iota.$$

A 1–1 mapping of a finite set onto itself is called a *permutation*. The usual composition of mappings provides a binary operation for permutations on the same set. Furthermore, whenever a collection of permutations is closed with respect to this composition, Axioms 2, 3, and 4 are autumatically satisfied and it is called a *permutation group*. If a permutation group A acts on object set X, then $|A|$ is the *order* of this group and $|X|$ is its *degree*.

Whenever A and B are permutation groups acting on the sets X and Y, respectively, we will write $A \cong B$ to mean that A and B are *isomorphic groups*. However $A \equiv B$ indicates not only isomorphism but that A and B are *identical permutation groups*. More specifically, $A \cong B$ if there is a 1–1 map $h : A \leftrightarrow B$ between the permutations such that for all $\alpha, \beta \in A$, $h(\alpha\beta) = h(\alpha)h(\beta)$. To define $A \equiv B$ precisely, we also require another 1–1 map $f : X \leftrightarrow Y$ between the objects such that for all $x \in X$ and $\alpha \in A$, $f(\alpha x) = h(\alpha)f(x)$.

An *automorphism* of a graph G is an isomorphism of G with itself. Thus each automorphism α of G is a permutation on the node set V which preserves adjacency. Of course, α sends any node onto another of the same degree. Obviously any automorphism followed by another is also an automorphism, hence the automorphisms of G form a permutation group, $\Gamma(G)$, which acts on the nodes of G. It is known as *the group of* G. Note that the group of a graph and the group of its complement are identical: $\Gamma(\overline{G}) \equiv \Gamma(G)$.

The identity map from V onto V is of course always an automorphism of G. For some graphs, it is the only automorphism; these are called *identity graphs*. The smallest nontrivial identity tree has seven nodes and is shown in Figure 8.1, as is an identity graph with six nodes.

Operations on Permutation Groups

There are several important operations on permutation groups which produce other permutation groups. We now develop four such binary operations: sum, product, wreath product, and power group.

Let A be a permutation group of order $m = |A|$ and degree s acting on set $X = \{x_1, x_2, \ldots, x_s\}$, and let B be another permutation group of order $n = |B|$ and degree t acting on the set $Y = \{y_1, y_2, \ldots, y_t\}$. For example, consider $A = Z_3$, the cyclic group of degree 3, which acts on $X = \{1, 2, 3\}$. Then the three permutations of Z_3 may be written $(1)(2)(3)$, $(1\,2\,3)$, and $(1\,3\,2)$. When $B = S_2$, the symmetric group of degree 2, acting on $Y = \{a, b\}$, we have the permutations $(a)(b)$ and (ab). We will use these two permutation groups to illustrate the binary operations defined here.

Their *sum* (sometimes called product or direct product and denoted accordingly) $A + B$ is a permutation group which acts on the disjoint union $X \cup Y$, whose elements are all the ordered pairs, written $\alpha + \beta$, of permutations α in A and β in B, . Any object z of $X \cup Y$ is permuted by $\alpha + \beta$ according to the rule:

$$(\alpha + \beta)(z) = \begin{cases} \alpha z, & z \in X; \\ \beta z, & z \in Y. \end{cases}$$

Thus $Z_3 + S_2$ contains 6 permutations each of which can be written as the sum of permutations $\alpha \in Z_3$ and $\beta \in S_2$ such as $(1\,2\,3)(ab) = (1\,2\,3) + (ab)$.

The *product* (also known as cartesian product) $A \times B$ of A and B is a permutation group which acts on the set $X \times Y$ and whose permutations are all the ordered pairs, written $\alpha \times \beta$, of permutations α in A and β in B. The object (x, y) of $X \times Y$ is permuted by $\alpha \times \beta$ as expected:

$$(\alpha \times \beta)(x, y) = (\alpha x, \beta y).$$

The product $Z_3 \times S_2$ also has order 6 but while the degree of the sum $Z_3 + S_2$ is 5, that of the product is 6. The permutation in $Z_3 \times S_2$ corresponding to $(1\,2\,3)(ab)$ in the sum is $(1a\ 2b\ 3a\ 1b\ 2a\ 3b)$, where for brevity $1a$ denotes $(1, a)$, etc.

The *wreath product* $A[B]$ of "A around B" also acts on $X \times Y$. It is often called the *composition* "A of B." For each $\alpha \in A$ and any sequence $(\beta_1, \beta_2, \ldots, \beta_s)$ of s (not necessarily distinct) permutations in B, there is a unique permutation in $A[B]$ written $(\alpha; \beta_1, \beta_2, \ldots, \beta_s)$ such that for any (x_i, y_j) in $X \times Y$:

$$(8.1) \qquad (\alpha; \beta_1, \beta_2, \ldots, \beta_s)(x_i, y_j) = (\alpha x_i, \beta_i y_j).$$

Table 8.1 OPERATIONS ON PERMUTATION GROUPS

			Sum	Product	Wreath Product	Power
group	A	B	$A+B$	$A \times B$	$A[B]$	B^A
objects	X	Y	$X \cup Y$	$X \times Y$	$X \times Y$	Y^X
order	m	n	mn	mn	mn^s	mn
degree	s	t	$s+t$	st	st	t^s

The wreath product $Z_3[S_2]$ has degree 6 but its order is 24. Each permutation in $Z_3[S_2]$ may be written in the form in which it acts on $X \times Y$. Using the same notation $1a$ for the ordered pair $(1,a)$ and applying the definition (8.1), one can verify that $((1\,2\,3);(a)(b),(ab),(a)(b))$ is expressible as $(1a\ 2a\ 3b\ 1b\ 2b\ 3a)$. Note that $S_2[Z_3]$ has order 18 and so is not isomorphic to $Z_3[S_2]$.

The *power group* denoted by B^A acts on Y^X, the set of all functions from X into Y. We will always assume that the power group acts on more than one function. For each pair of permutations α in A and β in B there is a unique permutation, written β^α in B^A. We specify the action of β^α on any function f in Y^X by the following equation which gives the image of each $x \in X$ under the function $\beta^\alpha f$:

$$(8.2) \qquad\qquad (\beta^\alpha f)(x) = \beta f(\alpha x).$$

The power group $S_2^{Z_3}$ has order 6 and degree 8. It is easy to see by applying (8.2) that the permutation in this group obtained from $\alpha = (1\,2\,3)$ and $\beta = (ab)$ has one cycle of length 2 and one of length 6.

Table 8.1 summarizes the information concerning the order and degree of each of these four operations.

We now see that three of these operations are not all that different.

Theorem 8.1 The three groups $A+B$, $A \times B$, and B^A are isomorphic.

Proof It is easy to show that $A+B \cong A \times B$. To see that $A+B \cong B^A$, we define the map $f : B^A \to A+B$ by $f(\alpha;\beta) = \alpha^{-1}\beta$, and verify that f is an isomorphism. Note that these three operations are commutative; in fact, $A+B \equiv B+A$, $A \times B \equiv B \times A$, and $B^A \cong A^B$. $\square$

Table 8.2 PERMUTATION GROUPS OF DEGREE P

Group	Symbol	Order	Definition
Symmetric	S_p	$p!$	All permutations on $\{1, 2, \ldots, p\}$
Alternating	A_p	$p!/2$	All even permutations on $\{1, 2, \ldots, p\}$
Cyclic	Z_p	p	Generated by $(1\,2\,\cdots\,p)$
Dihedral	D_p	$2p$	Generated by $(1\,2\,\cdots\,p)$ and $(1\,p)(2\,p-1)\cdots$
Identity	E_p	1	$(1)(2)\cdots(p)$ is the only permutation.

Table 8.2 introduces notation for five well-known permutation groups of degree p. In these terms, we can describe the groups of two familiar graphs with p nodes.

Theorem 8.2   The group $\Gamma(G)$ is S_p if and only if G is K_p or $\overline{K}_p$. The group $\Gamma(G)$ is D_p if and only if G is a cycle of length p or its complement.   □

Thus two particular groups of degree p, namely S_p and D_p, belong to graphs with p nodes. For all $p \geq 6$, there exists an identity graph with p nodes, and in fact whenever $p \geq 7$, there is an identity tree.

Graphs with a Given Group

König [K4,p.5] asked: When is a given abstract group isomorphic with the group of some graph? An affirmative answer to this question was given constructively by Frucht [F11]. His proof that every group is the group of some graph makes use of the Cayley "color-digraph of a group" [C5] which we now define. Let $F = \{f_0, f_1, f_2, \ldots, f_n - 1\}$ be a finite group of order n whose identity element is f_0. Let each nonidentity element have associated with it a different color. The *color-digraph* of F, denoted $D(F)$, is a complete symmetric digraph whose nodes are the n elements of F. In addition, each arc of $D(F)$, say from f_i to f_j, is labeled with a color associated with the element $f_i^{-1} f_j$ of F. Of course, in practice we simply label both nodes and arcs of $D(F)$ with the elements of F.

For example, consider the cyclic group of order 3, $Z_3 = \{0, 1, 2\}$. The color-digraph $D(Z_3)$ is shown in Figure 8.2.

Frucht discovered the next result which is simple but very useful.

Theorem 8.3   Every finite group F is isomorphic with the group of those automorphisms of $D(F)$ which preserve arc colors.

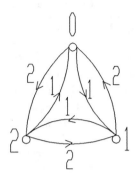

Figure 8.2   The color digraph of the cyclic group.

Proof   To construct a graph G whose group $\Gamma(G)$ is isomorphic with F, Frucht replaced each arc $f_i f_j$ in F by a doubly rooted graph. This is done in such a way that every arc of the same color is replaced by the same graph. Suppose that $f_i f_j = f_k$. Then the graph replacing arc $f_i f_j$ is a tree consisting of the path $f_i u_i v_i f_j$ with a path of length $2k - 3$ from u_i and a path of length $2k - 2$ from v_i. The resulting graph G has $n^2(2n - 1)$ nodes and $\Gamma(G) \cong F$. $\square$

Theorem 8.4   For every finite abstract group F, there exists a graph G such that $\Gamma(G)$ and F are isomorphic. $\square$

The graph obtained by this method from the cyclic group Z_3 is shown in Figure 8.3a. It should be clear from this example that the number of nodes in any graph so constructed may be excessive. When the group is known to have $m < n$ generators, a smaller graph can be obtained by modifying the color-graph to only include directed edges which correspond to the m generators. Thus a graph containing $n(m + 1)(2m + 1)$ nodes can be obtained for the given group. Since Z_3 can be generated by one element, there is a graph with 18 nodes for Z_3. It is shown in Figure 8.3b.

The inefficiency of even this improvement of the method of construction was illustrated in [HP1] where it was shown that the unique smallest graph whose automorphism group is Z_3 has 9 nodes and 15 edges.

Later Frucht [F12] showed that one could also specify that G be cubic. Sabidussi [S1] then showed that there are many graphs with a given abstract group having one of several other specified properties. Babai [B1] gives an excellent survey of further work in this area.

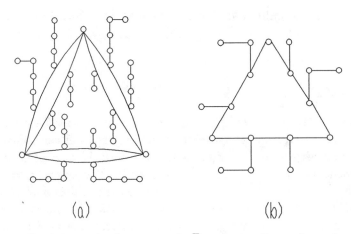

Figure 8.3   Frucht's graph whose group is Z_3 and a smaller such graph.

EXERCISES 8.1

1. If G is a connected graph, then $\Gamma(nG) = S_n[\Gamma(G)]$.

      (Frucht [F13])

2. If G_1 and G_2 are disjoint, connected, nonisomorphic graphs, then $\Gamma(G_1 \cup G_2) \equiv \Gamma(G_1) + \Gamma(G_2)$.

3. $\Gamma(G_1 \cup G_2) \equiv \Gamma(G_1) + \Gamma(G_2)$ if and only if no component of G_1 is isomorphic with a component of G_2.

4. $\Gamma(G_1 + G_2) \equiv \Gamma(G_1) + \Gamma(G_2)$ if and only if no component of $\overline{G}_1$ is isomorphic with a component of $\overline{G}_2$.

5. The group of the corona $G_1 \circ G_2$ is

$$\Gamma(G_1 \circ G_2) \equiv \Gamma(G_1)[E_1 + \Gamma(G_2)]$$

   if and only if G_1 or $\overline{G}_2$ has no isolated nodes.

      (Frucht and Harary [FH1])

6. Draw the unique smallest graph whose automorphism group is C_3. It has 9 nodes and 15 edges.   (Harary and Palmer [HP1])

7. Determine the group of each connected graph with four or fewer nodes.

8. Express the groups of the following graphs in terms of operations on familiar permutation groups:
   a. $\overline{3K}_2$ b. $\overline{K}_2 + C_4$ c. $K_{m,n}$ d. $K_4 \cup C_4$.

9. There are no nontrivial identity graphs with less than 6 nodes.

10.   There are no cubic identity graphs with less than 12 nodes.

11.   Construct a cubic graph whose group is cyclic of order 3.

8.2   SYMMETRIC GRAPHS

A systematic study of symmetry in graphs was initiated by Foster [F6], who made a tabulation of symmetric cubic graphs. Two nodes u and v of the graph G are *similar* if for some automorphism α of G, $\alpha(u) = v$. A *fixed node* is not similar to any other node. Two edges $x_1 = u_1v_1$ and $x_2 = u_2v_2$ are called *similar* if there is an automorphism α of G such that $\alpha(\{u_1, v_1\}) = \{u_2, v_2\}$. We consider only graphs with no isolated nodes. A graph is *node-symmetric* if every pair of nodes are similar; it is *edge-symmetric* if every pair of edges are similar; and it is *symmetric* if it is both node-symmetric and edge-symmetric. The smallest graphs that are node-symmetric but not edge-symmetric (the triangular prism $K_3 \times K_2$) and vice versa (the star $K_{1,2}$) are shown in Figure 8.4.

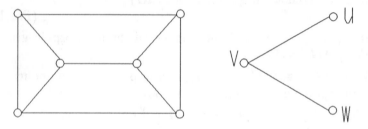

Figure 8.4   A node-symmetric and an edge-symmetric graph.

Note that if α is an automorphism of G, then it is clear that $G - u$ and $G - \alpha(u)$ are isomorphic. Therefore, if u and v are similar, then $G - u \cong G - v$. Surprisingly, the converse of this statement is not true. The graph in Figure 8.5 provides a counterexample, see [HP2]. It is the smallest graph which has dissimilar nodes u and v such that $G-u \cong G-v$.

The *degree of an edge* $x = u_1u_2$ is the unordered pair d_1, d_2 with $d_i = \deg u_i$, $i = 1, 2$. A graph is *edge-regular* if all edges have the same degree. When $m \neq n$, the complete bipartite graphs $K_{m,n}$ are edge-symmetric but not node-symmetric and are edge-regular of degree m, n.

We next state a theorem due to E. Dauber whose corollaries describe properties of edge-symmetric graphs. Note the obvious but important observation that every edge-symmetric graph is edge-regular.

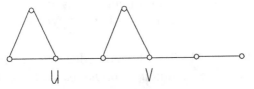

Figure 8.5   A counterexample to a conjecture.

Theorem 8.5   Every edge-symmetric graph with no isolated nodes is node-symmetric or bipartite.

Proof   Consider an edge-symmetric graph G with no isolated nodes, having q edges. Then for any edge x, there are at least q automorphisms $\alpha_1, \alpha_2, \ldots, \alpha_q$ of G which map x onto the edges of G. Let $x = v_1 v_2$, $V_1 = \{\alpha_1(v_1), \alpha_2(v_1), \ldots, \alpha_q(v_1)\}$, and $V_2 = \{\alpha_1(v_2), \alpha_2(v_2), \ldots, \alpha_q(v_2)\}$. Since G has no isolated nodes, the union of V_1 and V_2 is V. There are two possibilities: V_1 and V_2 are disjoint or they are not.

Case 1. If V_1 and V_2 are disjoint, then G is bipartite.

Consider any two nodes u_1 and w_1 in V_1. If they are adjacent, then there is an edge y joining them. Hence for some automorphism α_i, we have $\alpha_i(x) = y$. This implies that one of these two nodes is in V_1 and the other is in V_2, a contradiction. Hence V_1 and V_2 constitute a partition of V such that no edge joins two nodes in the same subset. By definition, G is bipartite.

Case 2. If V_1 and V_2 are not disjoint, then G is node-symmetric.

Let u and w be any two nodes of G. We wish to show that u and w are similar. If u and w are both in the same set, say V_1, then there exists an automorphism α with $\alpha(v_1) = u$ and β with $\beta(v_1) = w$. Thus $\beta\alpha^{-1}(u) = w$ so that any two nodes u and w in the same subset are similar. If $u \in V_1$ and $w \in V_2$, let v be a node in $V_1 \cap V_2$. Since v is similar with u and with v, u and v are similar to each other.   □

Corollary 8.5a   If G is edge-symmetric and the degree of every edge is d_1, d_2 with $d_1 \neq d_2$, then G is bipartite.   □

Corollary 8.5b   If a graph G with no isolated nodes is edge-symmetric, has an odd number of nodes, and the degree of every edge is d_1, d_2 with $d_1 = d_2$, then G is node-symmetric.   □

Corollary 8.5c If G is edge-symmetric, has an even number of nodes, and is k-regular with $k \geq p/2$, then G is node-symmetric. □

With these three corollaries, the only edge-symmetric graphs not yet characterized have an even number of nodes and are k-regular with $k < p/2$.

n-Transitive Graphs

Following Tutte [T16], an *n-route* is a walk of length n with specified initial node in which no edge succeeds itself. A graph G is *n-transitive*, $n \geq 1$, if it has an n-route and there is always an automorphism of G sending each n-route onto any other n-route. Obviously, a cycle of any length is n-transitive for all n, and a path of length n is n-transitive. Note that not every edge-symmetric graph is 1-transitive. For example, in the edge-symmetric graph $P_3 = v_1v_2v_3$, there is no automorphism sending the 1-route v_1v_2 onto the 1-route v_2v_3.

If W is an n-route $v_0v_1v_2 \cdots v_n$ and u is any node other than v_{n-1} adjacent with v_n, then the n-route $v_1v_2 \cdots v_n u$ is called a *successor* of W. If W terminates in an endnode of G, then obviously W has no successor. For this reason, it is specified in the next two theorems that G is a graph with no endnodes. We now have a sufficient condition [T16, p.60] for n-transitivity.

Theorem 8.6 Let G be a connected graph with no endnodes. If W is an n-route such that there is an automorphism of G from W onto each of its successors, then G is n-transitive. □

There is a simple relationship [T16, p.61] between n-transitivity and the girth of a graph.

Theorem 8.7 If G is connected, n-transitive, is not a cycle, has no endnodes and has girth g, then $n \leq 1 + g/2$. □

Using Theorem 8.6, it can be shown that the Heawood graph in Figure 8.6 is 4-transitive. Furthermore, it is easily seen from Theorem 8.7 that this graph is not 5-transitive. It is cubic and has girth 6.

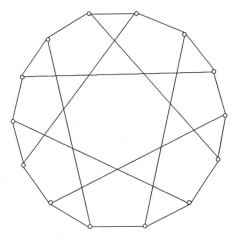

Figure 8.6   The Heawood graph.

Cages

There are regular graphs called "cages" which are, in a sense, even more highly symmetric than n-transitive graphs. An *n-cage*, $n \geq 3$, is a cubic graph of girth n with the minimum possible number of nodes. This class of graphs has been generalized to (r, n)-*cages*, which are r-regular graphs with girth n having minimum order. Thus an n-cage is equivalent to a $(3, n)$-cage. The unique $(2, n)$-cages are the cycles C_n. It was shown (see [T16, pp.71-83]) that there are $(3, n)$-cages for all $n \geq 3$, and that for $n = 3$ to 8, such $(3, n)$-cages are unique. It is interesting to note that not only are these cages n-transitive, but for each pair of n-routes in such a cage, there is a unique automorphism which maps one onto the other. The $(3, 8)$-cage is illustrated in Figure 8.7.

It is easy to see that the $(r, 3)$-cages are uniquely K_{r+1}. Erdös and Sachs [ES2] showed that (r, n)-cages always exist.

Theorem 8.8   For all pairs of integers $r, n \geq 3$, there is an (r, n)-cage and its order is bounded by

$$\left(\frac{r-1}{r-2}\right)[(r-1)^{n-1} + (r-1)^{n-2} + (r-4)].\qquad \square$$

The comprehensive survey article by Wong [W6] presents additional results on (r, n)-cages, as does the paper [HK1].

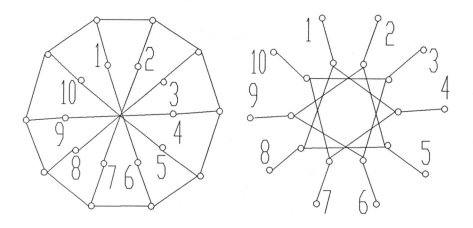

Figure 8.7 The $(3,8)$-cage is the union of the above graphs as labeled.

EXERCISES 8.2

1. Every connected node-symmetric graph is a block.

2. The $(r,4)$-cages are uniquely $K_{r,r}$.   (Chartrand and Lesniak [CL1])

3. Graph G is a *circulant* if it contains a spanning cycle $v_0 v_1 \cdots v_{p-1} v_0$ such that whenever the edge $v_i v_{m+n}$ is in G, so are all the edges $v_i v_j$ where $j - i \equiv n \pmod{p}$. A connected graph with a prime number p of nodes is node-symmetric if and only if it is a circulant.

    (Turner [T11])

4. If G is a connected graph with at least one cycle, then the girth $g(G)$ satisfies $g(G) \le 2d(G) + 1$.

5. Prove or disprove the following eight statements: If two graphs are node-symmetric (edge-symmetric), then so are their join, product, composition, and corona.

8.3   DISTANCE SYMMETRY

Some properties concerning symmetry in graphs refer directly to distance between nodes. Actually, edge-symmetry could be considered in this category. A graph G is edge-symmetric if for every two pairs of nodes at distance one from one another there exists an automorphism of G which maps one pair of nodes into the other pair. In this section we examine two classes of graphs whose distance symmetry conditions are more restrictive.

Distance-Transitive Graphs

A graph G is *distance-transitive* if for every set of four nodes u, v, w, z in G for which $d(u, v) = d(w, z)$, there is an automorphism α such that $\alpha(u) = w$ and $\alpha(v) = z$. It is clear from the definition that any distance-transitive graph is both node-symmetric and edge-symmetric and therefore symmetric. However, the converse is not true.

For $n \geq 5$ relatively prime to $\lceil (n-1)/2 \rceil$, the *generalized Petersen graph* of order $2n$ is constructed from two copies of C_n. Label the consecutive nodes of one cycle by $u_0, u_1, u_2, \ldots, u_{n-1}$ and those of the other by $v_0, v_1, v_2, \ldots, v_{n-1}$. Then join each node u_i to $v_{(i+\lceil (n-1)/2 \rceil) \bmod n}$. All generalized Petersen graphs are symmetric. Figure 8.8 displays a graph which is symmetric but not distance-transitive. The graph is symmetric since it is a generalized Petersen graph. To see that this graph is not distance-transitive, consider the 4 nodes v_1, v_4, v_2, v_{12}. For these nodes, $d(v_1, v_4) = d(v_2, v_{12}) = 3$. However, there can be no automorphism which maps v_1 to v_2 and v_4 to v_{12}, because v_1 and v_4 are joined by two geodesics, while v_2 and v_{12} are joined by three.

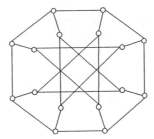

Figure 8.8   A symmetric graph which is not distance-transitive.

Distance-transitive graphs are regular, so one might organize an examination of these graphs according to degree. Biggs and Smith [BS4] showed that there are exactly twelve cubic distance-transitive graphs, and Smith [S23] showed that there are fifteen distance-transitive 4-regular graphs. For a node $v \in G$, the ith *neighborhood of* v is $N_i(v) = \{u \in G : d(u, v) = i\}$, and as above $d_i(v) = |N_i(v)|$. A useful result about distance-transitive graphs is the following.

Theorem 8.9   Let u and v be nodes at distance $d(u, v) = i$ in a distance-transitive graph G. Call $|N_1(u) \cap N_i(v)| = a_i$, $|N_1(u) \cap N_{i+1}(v)| = b_i$, and $|N_1(u) \cap N_{i-1}(v)| = c_i$. Then these numbers depend only on i and are independent of the choice of u and v. $\qquad\qquad \square$

Biggs [B8,p.135] observed that for a distance-transitive graph G which is necessarily regular, say of degree k, $a_0 = 0$, $b_0 = k$, and $c_1 = 1$. He then went on to prove the following.

Theorem 8.10 Let G be a distance-transitive graph with parameters a_i, b_1, and c_i as described above. Then

1. $d_{i-1}(v)b_{i-1} = d_i(v)c_i$ $(1 \leq i \leq d)$.
2. $1 \leq c_2 \leq c_3 \leq \cdots \leq c_d$.
3. $k \geq b_1 \geq b_2 \geq \cdots \geq b_{d-1}$.   □

Note that the definition of a distance-transitive graphs requires that for any two pairs of nodes whose distances apart match, there is an automorphism which sends one pair into the other. Rather than consider pairs, one could consider a corresponding problem for n-tuples. A graph G is *n-tuple distance-transitive* if for all sequences $(u_1, u_2, \ldots, u_n)$ and $(v_1, v_2, \ldots, v_n)$ with $d(u_i, u_j) = d(v_i, v_j)$ for all i, j, there is an automorphism α of G such that $\alpha(u_i) = v_i$. Thus distance-transitive graphs are just 2-tuple distance-transitive graphs. Cameron [C1] characterized 6-tuple distance-transitive graphs and showed that these graphs are also n-tuple distance-transitive for all n.

Theorem 8.11 If G is a 6-tuple distance-transitive graph, then G is one of the following:

1. $K(t, t, \ldots, t)$, the complete k-partite graphs with each part having t nodes.
2. $K_{k+1,k+1}$ minus a 1-factor.
3. C_p.
4. The line graph $L(K_{3,3})$.
5. The icosahedron.
6. The 9-regular graph on 20 nodes whose nodes correspond to the 3-subsets of a 6-set with two nodes adjacent whenever their intersection is a 2-set.   □

Distance-Regular Graphs

Let G be a k-regular graph with diameter d. Then G is *distance-regular* if there are positive integers $b_0 = k, b_1, b_2, \ldots, b_d - 1, c_1 = 1, c_2, c_3, \ldots, c_d$, such that for each pair of nodes u, v at distance j apart, we have

(8.3) The number of nodes at distance $j - 1$ from v which are neighbors of u is c_j $(1 \leq j \leq d)$; and

(8.4) The number of nodes at distance $j + 1$ from v which are neighbors of u is b_j $(0 \leq j \leq d - 1)$.

By Theorem 8.9 every distance-transitive graph is distance-regular. Thus numerous familiar graphs, such as cycles, complete graphs, hypercubes, and regular complete bipartite graphs are distance-regular. Although distance-regularity places no restrictions on the automorphism group of G, most distance-regular graphs are also distance-transitive. However, as Biggs noted [B8,p.139] with the following example due to Adel'son-Velski [A1], not all are. Let G be the graph with 26 nodes v_i, w_i, $0 \leq i \leq 12$, where adjacency is defined as follows: v_i and v_j are adjacent when $i - j \equiv 1, 3, 4, 9, 10, 12$ (mod 12); w_i and w_j are adjacent when $i - j \equiv 2, 5, 6, 7, 8, 11$ (mod 12); and v_i and w_j are adjacent when $i - j \equiv 0, 1, 3, 9$ (mod 12). Then G is distance-regular, but since it has no automorphism mapping a_i to b_j, it is not distance-transitive. Other examples of distance-regular graphs which are not distance-transitive were found by Weisfeiler [W1].

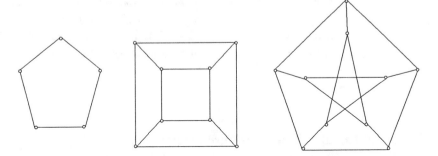

Figure 8.9   Some familiar distance-regular graphs.

For a distance-regular graph G, Smith [S22] established the following relationships for its parameters.

Theorem 8.12   If G is a distance-regular graph with parameters c_j and b_j, then

$$1 = c_1 \leq c_2 \leq \cdots \leq c_d \qquad \text{and} \qquad k = b_0 \geq b_1 \geq \cdots \geq b_{d-1}.$$

Proof Let v be a node of G and $w \in N_{j+1}(v)$ with $0 < j < d$. Then $|N_1(v) \cap N_j(w)| = c_{j+1} \neq 0$. Hence, there is a node u in $N_1(v) \cap N_j(w)$. Then $N_1(w) \cap N_{j-1}(u) \subset N_1(w) \cap N_j(v)$, so $c_j \leq c_{j+1}$. The argument for the parameters b_j is analogous.   □

Additional relationships between the parameters of a distance-regular graph were obtained recently by Nomura [N5]. The classification of cubic distance-regular graphs was completed by Biggs, Boshier and Shawe-Taylor [BBS1]. Just one such graph is not distance-transitive; it has 126 nodes (see Biggs [B8, p.164]).

Taylor and Levingston [TL1] studied distance-regular graphs of small diameter by considering the sizes $d_i(v)$ of the sets $N_i(v)$. For diameters 3 and 4, they characterized such graphs when $d_1 = d_2$. To describe these, they use a special graph and a construction on it. A *regular 2-graph* (W, S) consists of a set W together with a set S of 3- subsets of W such that every 4-subset of W contains an even number of element of S and every 2-subset of W is contained in the same number of elements of S. For results on these graphs see Taylor [3].

For the construction, take $V(G) = W \cup W'$, where W' is a second copy of W and has prime-labeled nodes. Define adjacency in G as follows. Choose some node $v \in W$ and join it to all other nodes in W. For each pair of nodes $u, w \neq v$ join u and w if $\{u, v, w\} \in S$, otherwise join u to u' and w to w'. Then join u' to w' whenever u is joined to w. Graph G is distance-regular as long as S does contain *all* 3-subsets of W. This is called the "doubling construction". We can now state the characterizations given by Taylor and Levingston [TL1].

Theorem 8.13 Let G be a distance-regular graph.

1. If G has diameter 3, then $d_1 = d_2$ if and only if g is the "doubled-graph" of a regular 2-graph or $G = C_7$.

2. If the diameter of G is at least 4, then $d_1 = d_2$ if and only if G is a cycle.   □

A graph G is *distance regularized* if, for any integer k and any nodes u and v, the number of neighbors of v which are at distance k from u only depends on $d(u, v)$ and node u. Bipartite distance regularized graphs in which nodes in the same part have the same parameters b_i, c_i are called *distance-biregular*. Godsil and Shawe-Taylor [GS1] categorized distance regularized graphs.

Theorem 8.14   Any distance regularized graph is either distance-regular or distance-biregular.   □

Strongly Regular Graphs

Bose [B20] introduced a class of graphs related to design theory and, like distance-regular graphs, defined in terms of neighborhoods. A k-regular graph $G \neq K_p$ or $\overline{K}_p$ is *strongly regular* if every pair of adjacent nodes has λ common neighbors and every pair of nonadjacent nodes has μ common neighbors. The numbers p, k, λ, μ are the parameters for a strongly regular graph. It is easy to verify that two of the graphs in Figure 8.9 are strongly regular. Which one is not? Since all pairs of nonadjacent nodes in a strongly regular graph must have the same number of common neighbors, all connected strongly regular graphs have diameter 2. Hence the hypercube Q_3 is not strongly regular.

There is a basic relationship between the parameters of a strongly regular graph.

Theorem 8.15   If G is a strongly regular graph with parameters p, k, λ, μ, then $(p - k - 1)\mu = k(k - 1 - \lambda)$.

Proof   Let $N(v)$ and $M(v)$ be the sets of neighbors and nonneighbors, respectively, of node $v \in G$. Set $N(v)$ has k elements and $M(v)$ has $p - k - 1$. For each $u \in N(v)$, u and v have λ common neighbors. Each such node u has degree k and $k - 1 - \lambda$ neighbors in $M(v)$. Since there are k such nodes u, there are $k(k - 1 - \lambda)$ edges between $N(v)$ and $M(v)$.

Another way of counting the number of edges between $N(v)$ and $M(v)$ is to begin with a node $w \in M(v)$. Since nodes v and w are not adjacent, they have μ common neighbors, all of which are in $N(v)$. This is true for each of the $p - k - 1$ nodes in $M(v)$. Therefore there are $(p - k - 1)\mu$ edges between $N(v)$ and $M(v)$. Hence $(p - k - 1)\mu = k(k - 1 - \lambda)$ □

Many strongly regular graphs are completely determined up to isomorphism by their parameters. See for example, Gewirtz [G3] and Seidel [S6]. For an excellent introduction to the theory of strongly regular graphs, the reader is referred to Cameron and Van Lint [CV1] and the survey article by Seidel [S7]. Construction of strongly regular graphs are given in Hubaut [H28].

EXERCISES 8.3

1. The hypercubes Q_n are distance-transitive.
2. The Petersen graph is distance-regular, and so is its line graph.
3. A graph G of diameter k is *antipodal* if $\{v \cup N_d(v) : v \in V(G)\}$ forms a partition of $V(G)$. If G is distance transitive and the automorphism group of G act imprimitively on V, then G is antipodal or bipartite.
    (Smith [S22])
4. If G is distance-regular of diameter d with parameters b_j and c_j, then $i + j \leq d$ implies $c_i \leq b_j$.   (Taylor and Levingston [TL1])
5. Construct a graph which is distance-biregular but not regular.
6. Let A and B be the adjacency matrices of a graph and its complement. If G is strongly regular with parameters p, k, λ, μ, then $A^2 = kI = \lambda A + \mu B$.
7. If G is strongly regular, then so is its complement.

FURTHER RESULTS

A.E. Brouwer, A.M. Cohen, and A. Neumiaer are completing a monograph entitled *Distance Regular Graphs* to be published in the series Ergebnisse der Mathematik by Springer-Verlag.

CHAPTER 9

Distance Sequences

A *sequence for a graph* is simply an invariant which consists of a list of numbers rather than a single number. The advantage of studying and using a sequence is that it is often nearly as easy to calculate as a single numerical invariant yet it carries far more information about the graph it represents. In this chapter we discuss a number of distance related sequences for a graph, display their relation to one another as well as to various concepts in graph theory, and indicate their uses in applications.

9.1 THE ECCENTRIC SEQUENCE

A sequence S is *graphical* if there is a graph which realizes S. Before discussing the eccentric sequence, we present results on the only graph sequence which predated it.

The Degree Sequence

The first sequence studied for graphs was the degree sequence. An existential characterization of graphical degree sequences was given by Erdős and Gallai [EG1]. The following constructive characterization was found independently by Havel [H17] and later by Hakimi [H1].

Theorem 9.1   The sequence $D = (d_1, d_2, \ldots, d_p)$ with
$$p - 1 \geq d_1 \geq d_2 \geq \cdots \geq d_p$$
is a graphical degree sequence if and only if the modified sequence
$$D' = (d_2 - 1, d_3 - 1, \ldots, d_{d_1+1} - 1, d_{d_1+2}, \ldots, d_p)$$
is a graphical degree sequence.

Proof   If D' is a graphical degree sequence, then so is D, since from a graph with degree sequence D' one can construct a graph with degree sequence D by adding a new node adjacent to the nodes having degrees $d_2 - 1, d_3 - 1, \ldots, d_{d_1+1} - 1$.

Now let G be a graph with degree sequence D. If a node of degree d_1 is adjacent to nodes of degree d_k for $k = 2$ to $d_1 + 1$, then the removal of this node results in a graph with degree sequence D'.

Suppose that G has no such node. We will show that from G one can always get another graph with degree sequence D having such a node. We assume that the nodes in G are labeled so that deg $v_i = d_i$ and that v_1 is a node of degree d_1 for which the sum of the degrees of the adjacent nodes is maximum. Then there are nodes v_i and v_j with $d_i > d_j$ such that $v_1 v_j$ is an edge but $v_1 v_i$ is not. Since $d_i > d_j$, there must be some node v_k adjacent to v_i but not to v_j. Removal of the edges $v_1 v_j$ and $v_k v_i$ and the addition of $v_1 v_i$ and $v_k v_j$ results in another graph with degree sequence D. But in this new graph, the sum of the degrees of the nodes adjacent to v_1 is greater than before since v_1 is now adjacent to v_i rather than v_j. By repeating this edge-switching process a finite number of times, we obtain a graph with degree sequence D in which v_1 has the desired property. $\square$

The theorem gives an effective algorithm for the construction of a graph with a given degree sequence, if one exists. If none exists, the algorithm cannot be applied at some step.

Algorithm 9.1   The sequence $D = (d_1, d_2, \ldots, d_p)$ with $p - 1 \geq d_1 \geq d_2 \geq \cdots \geq d_p$ is a graphical degree sequence if and only if the following procedure results in a sequence with every term zero.

1.   Determine the modified sequence D' as described in Theorem 9.1.

2.   Reorder the terms of D' so that they are in nonincreasing order, and call the resulting sequence D_1.

3.   Determine the modified sequence D'' of D_1 as in step 1 and reorder D'' as in step 2; call the reordered sequence D_2.

4.   Continue as long as only nonnegative terms are obtained. $\square$

If a sequence at an intermediate stage of the algorithm is known to be a graphical degree sequence, stop, since D itself is then established to be one also. To illustrate Algorithm 9.1, we test the sequence

$$D = (5,5,3,3,2,2,2)$$
$$D' = (4,2,2,1,1,2)$$
$$D_1 = (4,2,2,2,1,1)$$
$$D'' = (1,1,1,0,1)$$
$$D_2 = (1,1,1,1,0)$$

Clearly, D_2 is a graphical degree sequence, so D is also. The graph so constructed is shown in Figure 9.1.

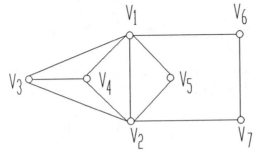

Figure 9.1 An example for Algorithm 9.1.

Eccentric Sequences

The *eccentric sequence* of a connected graph G is a list of the eccentricities of its nodes in nondecreasing order. Since there are often many nodes having the same eccentricity, we will simplify the sequence by listing it as

$$e_1^{m_1}, e_2^{m_2}, \ldots, e_k^{m_k},$$

where the e_i are the eccentricities ($e_i < e_{i+1}$) and m_i is the multiplicity of e_i. Thus, for the eccentric sequence $2,2,3,3,3,3,4,4$ for the graph of Figure 9.2, we will write $2^2, 3^4, 4^2$. If node u has eccentricity t and for some node v, $d(u,v) = t$, then v is called an *eccentric node* of u.

Some simple observations about the values e_i and m_i for a nontrivial connected graph are as follows:

1. Since for each pair of adjacent nodes u, v, and any third node w, $|d(u,w) - d(v,w)| \leq 1$, it follows that the e_i's are consecutive positive integers.

2. $\sum m_i = p$.

3. $e_1 = r(G)$ and $e_k = d(G)$, so $1 \le e_i \le p-1$.

4. Since there must be a pair of diametral nodes, $m_k \ge 2$.

5. Since the diameter is at most twice the radius, $e_k \le 2e_1$.

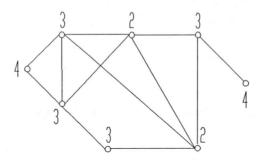

Figure 9.2   A graph and its eccentricities.

Lesniak [L3] showed that (possibly) except for the radius r, there are at least two nodes of each eccentricity between r and d.

Theorem 9.2   For $p \ge 2$, $m_i \ge 2$ except possibly for m_1.

Proof   Since the e_i's are consecutive positive integers, there is at least one node of eccentricity t for each integer t, $e_1 < t \le e_k$. Let u be a node with eccentricity $t > e_1$ in G, and let v be an eccentric node of u. Then $e(v) \ge t$. For a central node w, let P be a $v-w$ geodesic. Since $e(w) = e_1$, $d(v,w) \le e_1$. Since the eccentricities of adjacent nodes can differ by at most one, and $e(w) = e_1 < t \le e(v)$, some node x on P (possibly $x = v$) has eccentricity t. Since $d(u,v) = t > e_1 \ge d(x,v)$, node x must be distinct from u. Thus there are at least two nodes with eccentricity t.   □

Ostrand [O5] determined all nonisomorphic graphs of minimum order having specified radius and diameter. His result which follows will be useful for us.

Lemma 9.3   For all positive integers r and d satisfying $r \le d \le 2r-2$, there exist graphs with radius r and diameter d. The minimum order of such a graph is $r+d$. There are exactly $\lfloor (d-r)/2 \rfloor + 1$ nonisomorphic graphs of order $r+d$, radius r, and diameter d. Each graph consists of a path $u_0, u_1, u_2, \ldots, u_d$ and a path $u_s, v_1, v_2, \ldots, v_{r-1}, u_{s+r}$ with only the nodes u_s and u_{s+r} in common ($0 \le s \le \lfloor (d-r)/2 \rfloor$).   □

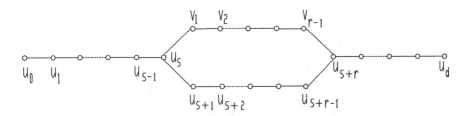

Figure 9.3   Minimum order graphs with given radius and diameter.

Theorem 9.3   Let $r^{m_1}, (r+1)^{m_2}, \ldots, (r+k-1)^{m_k}$ be the eccentric sequence of a graph on p nodes ($p \geq 2$). Then the following conditions must hold.

1. $k \leq r+1$, and $p \geq 2r+k-1$ for $k \leq r-1$ ($p \leq r+k-1$ otherwise).
2. $m_i \geq 2$ for $2 \leq i \leq k$.
3. $\sum m_i = p$.

Proof   Clearly condition 3 is necessary. The first part of condition 1 is equivalent to saying that the diameter is at most twice the radius. The second part of 1 follows from Lemma 9.3 for $d \leq 2r-2$, and the fact that for $d = 2r$ or $2r-1$, the path P_d is the minimal graph. Condition 2 must hold by Theorem 9.2.   □

Minimal Eccentric Sequences

We say that a sequence is *eccentric* if it is realizable as the eccentric sequence of some graph. Lesniak [L3] obtained the following characterization of eccentric sequences.

Theorem 9.4   A sequence S of positive integers is eccentric if and only if some subsequence T of S is eccentric.   □

One of the difficulties in using Theorem 9.4 is that the subsequence T may be the full sequence S itself. This led Nandakumar [N1] to consider the concept of minimal sequences. An eccentric sequence is *minimal* if it has no proper eccentric subsequences with the same number of distinct eccentricities. Nandakumar [N1] determined all minimal eccentric sequence with least eccentricity at most 2.

Theorem 9.5   There are exactly six minimal eccentric sequences with least eccentricity at most two, namely,
1^2; $1,2^2$; 2^4; $2^2,3^2$; $2,3^6$; and $2,3^2,4^2$.   □

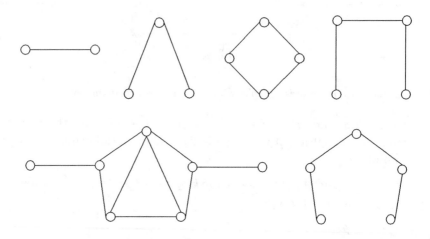

Figure 9.4   Graphs realizing the minimal eccentric sequences of Theorem 9.5.

EXERCISES 9.1

1. Which of the following are graphical degree sequences?

   a. $(4,3,3,3,2,2,2,1)$.

   b. $(8,6,5,4,3,2,2,1)$.

   c. $(5,5,5,3,3,3,3,3)$.

   d. $(5,4,3,2,1,1,1,1,1,1,1,1,1)$.

2. A sequence $(d_1, d_2, \ldots, d_p)$ where $\sum d_i = 2q$ is the degree sequence of a tree if and only if each d_i is a positive integer and $q = p - 1$.

3. Draw all graphs with degree sequence $(5,5,3,3,2,2)$.

4. The *eccentric mean* $\mu_e(G) = (1/p) \sum m_i e_i$. Determine $\mu_e(G)$ for each of the following.
   a. K_p b. P_p c. C_p d. $K_{a,b}$ e. $W_{1,n}$.

5. If G is a connected graph on p nodes, then $1 \le \mu_e(G) \le (3p-2)/4$ for even p and $1 \le \mu_e(G) \le (3p^2 - 2p - 1)/4p$ for odd p. The lower and upper bounds are achieved only when $G = K_p$ or P_p, respectively.

6. The conditions of Theorem 9.3 are sufficient to guarantee a graph realizing the given eccentric sequence when $r \leq 2$.

7. Verify that sequence $3^4, 4^3$ satisfies the conditions of Theorem 9.3, although it is not eccentric.

8. A finite nonempty set S of positive integers is an *eccentric set* if there exists a graph G all of whose eccentricities are elements of S. A nonempty set $S = \{e_1, e_2, \ldots, e_n\}$ of positive integers, listed in increasing order, is an eccentric set if and only if $n \leq e_1 + 1$ and the e_i are consecutive integers.   (Behzad and Simpson [BS2])

9. A spanning tree T of a connected graph G is *eccentricity preserving* if and only if it has the same eccentric sequence as G. Graph G has an eccentricity preserving spanning tree if and only if

   a. G is central and $d = 2r$, or
      G is bicentral and $d = 2r - 1$; and

   b. For each v with $e(v) > r(G)$, one of its neighbors u satisfies
      $e(u) = e(v) - 1$.   (Nandakumar [N1])

9.2   DISTANCE SEQUENCES

There are several graphical sequences that concern the distances between all pairs of nodes in a graph. We discuss these sequences in this section.

The Distance Degree Sequence

This sequence actually consists of a collection of sequences. For a node v in a connected graph G, let $d_i(v)$ be the number of nodes at distance i from v. The *distance degree sequence of node v* is

$$dds(v) = (d_0(v), d_1(v), d_2(v), \ldots, d_{e(v)}(v)).$$

Note the following:

(9.1) $d_0(v) = 1$ for all v; $d_1(v) = \deg v$.

(9.2)   The length of sequence $dds(v)$ is one more than the eccentricity of v.

(9.3) $\sum d_i(v) = p$.

The *distance degree sequence $dds(G)$ of a graph G* consists of the collection of sequences $dds(v)$ of its nodes, listed in numerical order. If a particular *dds* appears k times, we list it once with k as an exponent to indicate the multiplicity. For example, in Figure 9.5, $dds(t) = (1, 2, 1, 1)$, $dds(w) = (1, 3, 1)$, and $dds(G) = ((1, 1, 2, 1); (1, 2, 1, 1); (1, 3, 1)^3)$.

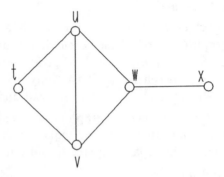

Figure 9.5   A graph to illustrate the distance degree sequence.

Distance degree sequences of graphs were studied by Randic [R1] for the purpose of distinguishing chemical isomers by their graph structure. Chemists have also proposed and discussed other sequences for this purpose. Their objective and hope is to develop a "chemical indicator" which would be useful both for the "graph isomorphism problem" and to predict various properties of the molecule at hand.

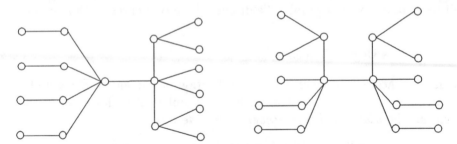

Figure 9.6   A pair of nonisomorphic trees with the same distance degree sequence.

Randic [R1] conjectured that a tree is determined by its distance degree sequence and verified it for all trees with 14 or fewer nodes. Slater [S20] disproved this conjecture by showing how to construct an infinite class of pairs of trees so that each pair has the same distance degree sequence. The smallest pair of graphs among his counterexamples are displayed in Figure 9.6.

For non-tree graphs, it is far easier to find nonisomorphic graphs with identical distance degree sequences. The smallest pair, with five nodes, are displayed in Figure 9.7. Each graph has *dds* equal to $((1,2,2)^3; (1,3,1)^2)$.

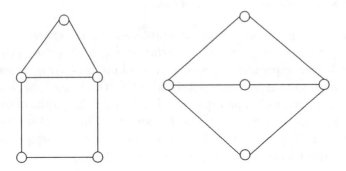

Figure 9.7   The smallest nonisomorphic graphs with the same distance degree

   sequence.

Bloom, Kennedy, and Quintas [BKQ1] were concerned with graphs in which all nodes have the same distance degree sequence. Such graphs are called *distance degree regular (DDR) graphs*. Thus a DDR graph has the property that $dds(G) = ((dds(v))^p)$, where v is any node in G. A DDR graph G is necessarily regular since $d_1(v) = d_1(w)$ for any two nodes v and w in G. However, the converse is not true. In fact, we have the following.

Theorem 9.6   Every regular graph containing a cutnode is not DDR.

Proof   Let G be a connected regular graph with a cutnode v, and let G_1 and G_2 be components of $G - v$. Suppose that an eccentric node of v in G lies in G_2, and let x be a neighbor of v which lies in G_1. Then the eccentricity of x within G is greater than the eccentricity of v in G. Thus $dds(x) \neq dds(v)$ by (9.2).   □

Bloom, Kennedy, and Quintas [BKQ1] showed, however, that each regular graph with diameter at most two is DDR. The following is a simple yet useful observation.

Theorem 9.7   If G is DDR, then G is self-centered and self-median.   □

Figure 9.8   A family of DDI graphs of order $p \geq 7$.

Now we turn our attention to graphs at the far extreme from DDR graphs. A connected graph G is *distance degree injective (DDI)* if the distance degree sequences of its nodes are all distinct. As opposed to DDR graphs, these graphs are completely asymmetric. Indeed, all DDI graphs have identity automorphism group. There are DDR graphs of every order and every diameter because of the K_n and C_n. On the other hand, using the graphs in Appendix 1 and the class of graphs in Figure 9.8, Bloom, Kennedy, and Quintas [BKQ2] showed the following.

Theorem 9.8   A smallest nontrivial DDI graph has order 7, and there exist DDI graphs for all orders $p \geq 7$.   □

Before the abbreviations used get too overwhelming, we list in Table 9.1 a dictionary of acronyms for this and the next section.

Table 9.1 **DICTIONARY OF ACRONYMS**

dds	distance degree sequence
DDR	distance degree regular
DDI	distance degree injective
ss	status sequence
SI	status injective
dd	distance distribution

The following result is from Bloom, Kennedy and Quintas [BKQ2].

Theorem 9.9   If G is a nontrivial graph for which both G and $\overline{G}$ are DDI, then both G and $\overline{G}$ have diameter 3.

Proof   The only graphs with diameter 1 are the complete graphs and for $p > 1$, they are DDR but not DDI. Next, suppose that G is a graph with diameter two. Then the distance degree sequence of any node v of G is $dds(v) = (1, \deg v, p - \deg v - 1)$. Since at least two nodes of G have the same degree, two nodes of G also have the same dds. Thus, no DDI graph has diameter 2. Finally, if $d(G) > 3$, then $d(\overline{G}) \leq 2$. Thus, a DDI graph cannot have diameter greater than three if its complement is also a DDI graph.   □

The Status Sequence

Recall that the *status* $s(v)$ *of a node* v in a connected graph G is the sum of the distances to all the other nodes in G. The *status sequence* $ss(G)$ of a connected graph G is the list of its status values arranged in nondecreasing order. The graph in Figure 9.9 has status sequence $(5^3, 7, 8)$.

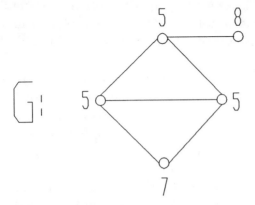

Figure 9.9   A graph to illustrate the status sequence.

The relationship between status sequences and median problems is analogous to the relationship between eccentric sequences and center problems. There are several properties which distinguish status sequences from eccentric sequences.

1.   The status values need not be consecutive integers.
2.   There need not be two nodes having maximum status.
3. $ss(G)$ is derivable from $dds(G)$: For the sequence
     $dds(v) = (d_0(v), d_1(v), \ldots, d_{e(v)}(v))$, we have

$$s(v) = \sum_{i=1}^{e(v)} i \cdot d_i(v).$$

It follows from item 3 that results about one sequence imply results about the other. For example, Slater's result showing the existence of two nonisomorphic graphs G_1 and G_2 for which $dds(G_1) = dds(G_2)$ yields a

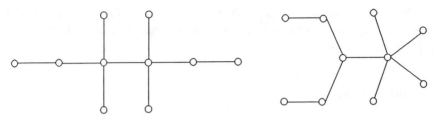

Figure 9.10   A pair of graphs with the same status sequence.

pair of nonisomorphic graphs with the same status sequence. Indeed, if $dds(G_1) = dds(G_2)$, then $ss(G_1) = ss(G_2)$. However, the converse is not true. We display in Figure 9.10 a pair of trees (due to Slater [S20]) whose status sequences are identical. It is easy to see that their distance degree sequences are different because their diameters differ as do their degree sequences.

Very little has been done with status sequences other than bounds on individual terms obtained by Entringer, Jackson, and Snyder [EJS1].

Obviously, all statuses of nodes in K_p are equal, as are those in C_p. Thus these graphs are *self-median* and as noted in Theorem 9.7, so are all DDR graphs. In searching for self-median graphs, one usually finds that the graph is regular. However, this need not be the case, as seen in an example provided to us by R.C. Entringer and displayed in Figure 9.11. There is at this time no nice characterization of self-median graphs.

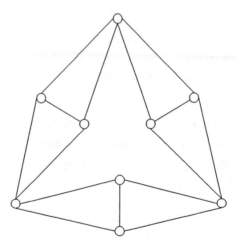

Figure 9.11   A nonregular self-median graph.

Self-median graphs might also be called *status sequence regular graphs*. Theorem 9.7 implies that these are a special subclass of DDR graphs. At the other extreme from self-median graphs, we may consider graphs G for which all of the terms of $ss(G)$ are distinct. We call these graphs *status injective (SI)*. Clearly, SI graphs are a special class of DDI graphs. Of the infinite class of DDI trees displayed in Figure 9.8, only the tree of order 7 is SI. This somewhat indicates the scarcity of SI graphs. An even more restrictive problem is that of finding nontrivial SI graphs whose statuses are consecutive integers. As yet, we have found no such graph.

EXERCISES 9.2

1. Determine the status sequences of K_p, $K_{a,b}$, and $W_{1,n}$.

2. The status sequence of P_p is

$$\begin{cases} (p^2-1)/4, ((p^2+4t^2-1)/4)^2 & \text{when } 1 \le t \le (p-1)/2, \, p \text{ odd}; \\ ((p^2+4t^2-4t)/4)^2 & \text{when } 1 \le t \le (p-2)/2, \, p \text{ even}. \end{cases}$$

3. The status sequence of C_p is

$$\begin{cases} (p^2-1)^p & p \text{ odd, and} \\ (p^2/4)^p & p \text{ even}. \end{cases}$$

4. For self-median graphs, $\Delta - \delta$ may be arbitrarily large.

5. Each connected regular graph of diameter at most two is DDR.
                                       (Bloom, Kennedy, and Quintas [BKQ1])

6. Construct a regular graph which is not DDR.

7. Every node-symmetric graph is DDR.
                                       (Bloom, Kennedy, and Quintas [BKQ1])

8. Verify that the graphs in Figure 9.7 have
   $dds(G) = ((1,2,2)^3; (1,3,1)^2)$.

9. Construct a DDR graph which is not edge-symmetric.

10. For each integer $r \ge 3$, the smallest order p for which there is a pair of nonisomorphic r-regular graphs having the same distance degree sequence is $p = r + 3$.   (Quintas and Slater [QS1])

11. There exists a nontrivial DDI graph with diameter k for all $k \ge 3$.
                                       (Bloom, Kennedy, and Quintas [BKQ2])

12. Every DDI graph has identity automorphism group.

13. If G has two nodes u and v with the same degree such that
    $e(u) = e(v) = 2$, then G is not DDI.

9.3   THE DISTANCE DISTRIBUTION

Let D_i be the number of pairs of nodes at distance i from one another in the connected graph G with diameter d. Then the *distance distribution* of G is the sequence

$$dd(G) = (D_1, D_2, \ldots, D_d).$$

Obviously, $dd(G)$ is obtained at once from from $dds(G)$ as $2D_i = \sum d_i(v)$ with the sum taken over all nodes v of G. Also, note that $D_1 = q$, the number of edges in G.

Although $dd(G)$ can be derived from $dds(G)$, it still contains a wealth of information and deserves a separate treatment. In fact, in certain problems, $dds(G)$ contains *too much* information and is cumbersome to work with; whereas $dd(G)$ is ideal for the problem. However, remembering that $dd(G)$ can be derived from $dds(G)$ is useful. For example, Figure 9.6 provides an immediate example of a pair of nonisomorphic graphs which have the same distance distribution. The next result due to M. Capobianco is easy to prove by induction on p and can be used to test the statistical hypothesis that a graph is connected.

Theorem 9.10   If G is a connected graph on p nodes, then $D_1 + D_2 \geq 2p - 3$.   □

Theorem 9.11   When G is a tree, D_2 is given by the degree sequence:

$$D_2 = \sum_{i=1}^{p} \binom{\deg v_i}{2}.$$

Proof   Let $N(v)$ be the set of neighbors of v. Each pair of nodes in $N(v)$ are joined by a unique path, which necessarily passes through v. The term $\binom{\deg v_i}{2}$ counts the number of pairs of nodes that are at distance two, via v_i, from each other. By summing over all v_i, the result follows.   □

For a connected graph G, let $s_k(G)$ be the kth partial sum of $dd(G)$, that is,

$$s_k(G) = \sum_{i=1}^{k} D_i.$$

The next (unpublished) result of M.O. Albertson, D. Berman, and F. Buckley concerns these partial sums for trees.

Theorem 9.12 Let T be any tree on p nodes. Then $s_k(T) \geq s_k(P_p)$, and equality holds for all k if and only if $T = P_p$.

Proof (by induction). The result is clear for small p. Assume it is true for all $t < p$ and let T be any tree on p nodes. Let $d(G) = d$ and $e(v) = d$, that is, v is the endnode of a diametral path in T. By the inductive hypothesis, $s_k(T - v) \geq s_k(P_{p-1})$ for all k. By attaching an extra node to an endnode of P_{p-1} (and thus forming P_p), we increase each D_i of P_{p-1} by exactly 1. By reattaching v to $T - v$, we increase each D_i of $T - v$ by at least 1. By D_d, we have accumulated an increase of $p - 1$, since each node of T can be reached from v by a path of length d or less. Thus, $s_k(T) \geq s_k(P_p)$ for each k. If $T \neq P_p$ then $d(P_p) = p - 1$ and $d(T) < p - 1$. Therefore

$$s_{p-2}(T) = s_{p-1}(T) = s_{p-1}(P_p) > s_{p-2}(P_p).$$

Hence equality holds for all k if and only if $T = P_p$.   □

Corollary 9.12   For any connected graph of order p and diameter d, we have

$$D_1 + D_2 + \cdots + D_k \; \geq \; pk - \binom{k+1}{2}, \quad \text{for all } k \leq d. \qquad \square$$

Note that Corollary 9.12 generalizes Theorem 9.10.

Amin, Siegrist, and Slater [ASS1] studied the "pair-connected reliability," which relates to the expected number of pairs of nodes that remain connected in a graph G when each edge fails independently with some fixed probability q, $0 < q < 1$. They showed that for trees T_1 and T_2, if $s_k(T_1) \geq s_k(T_2)$ for all k, then T_1 is uniformly more reliable than T_2, which means that it is more reliable for any given value of the probability q of edge failure.

If a graph has a path of length k, then it has a path of each smaller length. This may lead one to feel that $dd(G)$ must be a nonincreasing sequence. Not only is this not the case, but $dd(G)$ need not even be "unimodal". A sequence S_i is *unimodal* if there is some k for which $S_i \leq S_{i+1}$ for $i < k$, and $S_j \geq S_{j+1}$ for $j \geq k$. In Figure 9.12a the mode is central; and in Figure 9.12b, the mode is the diameter and the sequence is not unimodal.

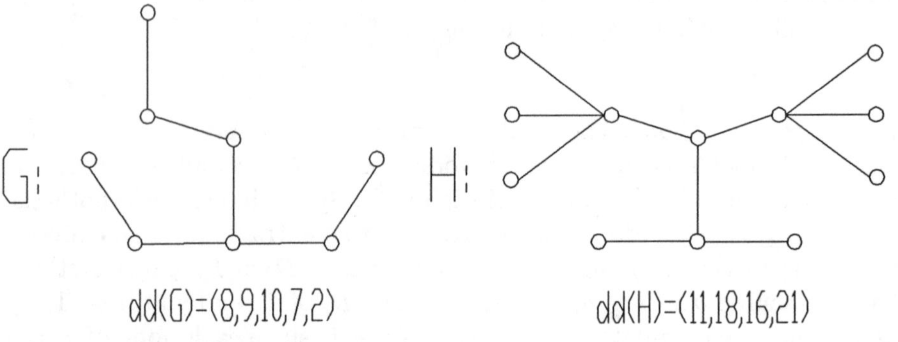

Figure 9.12   Two graphs and their distance distributions.

Uniform Distance Distributions

A distance distribution is *uniform* if $D_i = D_j$ for all i and j. Thus, if G has diameter d and $dd(G)$ is uniform, then $D_i = \binom{p}{2}/d$. Let U be the set of all graphs G for which $dd(G)$ is uniform. Clearly, $K_p \in U$ and $C_{2d+1} \in U$. Buckley and Superville [BS7] obtained the following result for uniform distance distributions.

Theorem 9.13

1. If $\binom{p}{2}$ is even, there exist at least two uniform distance distributions for p.

2. If $\binom{p}{2}$ is divisible by 3, there are at least two graphs of order p with uniform distance distributions, except for $p = 3$ and 6.

3. If $p = 12k + 2$ and $\binom{p}{2}$ is the product of two primes, then K_n is the unique graph of order p with a uniform distance distribution.   □

   In Table 9.2, we list all realizable uniform distance distributions for graphs with ten or fewer nodes. In the table, C_9^2 is the square of the cycle C_9, and graphs G^*, and H^* are displayed in Figure 9.13. Note that many graphs with uniform distance distributions can be constructed like H^*. That is, begin with the *double star* $S_{a,b} = \overline{K}_a + K_1 + K_1 + \overline{K}_b$ and then insert edges between certain nonadjacent (noncentral) nodes. Also, Let $C_{4k,2}$ be the graphs formed from C_{4k} as follows: Let the nodes of C_{4k} be labeled $1, 2, \ldots, 4k$; insert additional edges $(i, i + 2) \bmod 4k$, and delete edges $(1, 4k - 1)$ and $2k, 2k + 2$. For $p > 7$, the associated graphs in Table 9.2 are not necessarily unique.

Table 9.2 SOME UNIFORM DISTANCE DISTRIBUTIONS

n	Realizable Sequences	Associated Graphs
2	(1)	K_2
3	(3)	K_3
4	(6), $(3,3)$	K_4, $K_{1,3}$
5	(10), $(5,5)$	K_5, C_5
6	(15)	K_6
7	(21), $(7,7,7)$	K_7, C_7
8	(28), $(14,14)$	K_8, $C_{8,2}$
9	(36), $(18,18)$, $(12,12,12)$, and $(9,9,9,9)$	K_9, C_9^2, G^* and C_9
10	(45), $(15,15,15)$	K_{10}, H^*

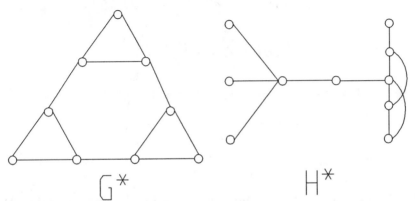

$$G^*$$ $$H^*$$

Figure 9.13   Two graphs for Table 9.2.

Mean Distance

The *mean distance* $\mu_D(G)$ of a connected graph G is the average of the distances between pairs of nodes in G. Of course, $\mu_D(G)$ can be calculated from $dd(G)$:

$$(9.4) \qquad \mu_D(G) = \sum_{i=1}^{d} iD_i \Big/ \binom{p}{2}.$$

Mean distance was examined by Doyle and Graver [DG1], who noted that it had been used by March and Steadman [MS1] in architecture as an aid in the evaluation of floor plans.

Since mean distance and the radius are both measures of the "central tendency" of a graph, Buckley and Superville [BS7] studied graphs G with the property that $\mu_D(G) = r(G)$. For such graphs we obtain a rather nice relationship between r and the D_i's from $dd(G)$. We know that

(9.5)
$$\sum_{i=1}^{d} D_i = \binom{p}{2}.$$

Knowing this and setting $\mu_D(G) = r(G)$, we obtain our result. Equation (9.4) says that $\mu_D(G) = \sum_{i=1}^{d} iD_i/\binom{p}{2}$, so

$$\mu_D(G) = \left(D_1 + \sum_{i=2}^{d} iD_i\right)/\binom{p}{2}$$

$$= \left(\binom{p}{2} - \sum_{i=2}^{d} D_i + \sum_{i=2}^{d} iD_i\right)/\binom{p}{2}$$

$$= \left(\binom{p}{2} + \sum_{i=2}^{d}(i-1)D_i\right)/\binom{p}{2}.$$

Upon setting this last expression equal to r and simplifying, we get

(9.6)
$$\sum_{i=2}^{d}(i-1)D_i = (r-1)\binom{p}{2}.$$

Recall that a self-centered graph has the property that all nodes have the same eccentricity. Thus, for a self-centered graph, $r(G) = d(G)$. The only self-centered graphs G satisfying $\mu_D(G) = r(G)$ are the complete graphs, as the following theorem indicates.

Theorem 9.14   If $r(G) = d(G) \geq 2$, then $\mu_D(G) \neq r(G)$. $\square$

We can now give a result of Buckley and Superville [BS7].

Theorem 9.15   If $d(G) = 3$, then $\mu_D(G) = r(G)$ if and only if $r(G) = 2$ and $D_1(G) = D_3(G)$.

Proof   If $d = 3$, then $r = 2$ or 3. Suppose that $\mu_D(G) = r(G)$. Then by Theorem 9.14, r must be 2. Substitution into (9.6) yields $D_2 + 2D_3 = \binom{p}{2}$. But (9.5) says that $D_1 + D_2 + D_3 = \binom{p}{2}$. Thus $D_1 = D_3$.

If $r(G) = 2$ and $D_1 = D_3$, then by substituting into (9.4) and using (9.5), we get

$$\mu_D(G) = \left(D_1 + D_2 + D_3\right) \Big/ \binom{p}{2}$$
$$= (2D_1 + 2D_2 + 2D_3)/(D_1 + D_2 + D_3) = 2 = r(G). \qquad \Box$$

Corollary 9.15   The only tree T of diameter three with $\mu_D(T) = r(T)$ is the tree having degree sequence $(4, 3, 1, 1, 1, 1, 1)$. $\qquad \Box$

A *caterpillar* is a tree T having a diametral path incident with every edge of T. A tree which we call the *double starred path* $P_{a;b,c}$ is the graph formed from P_a by attaching b pendant edges at one end and c pendant edges at the other. By joining various pairs of endnodes in such graphs, we were able to show that there are graphs for every diameter $d \neq 2$ for which $\mu_D(G) = r(G)$. F. Halberstam showed that there are, in fact, always caterpillars with this property.

The *edge density* $\rho(G)$ of (p,q)-graph G is $q/\binom{p}{2}$. Hendry [H20] showed that for each rational number $t > 1$ there are infinitely many graphs G with $\mu_D(G) = t$. In another paper [H21] he showed that for each rational t there are sparse graphs G with $\mu_D(G) = t \geq 2$ and dense graphs with $\mu_D(G) = t > 1$. Specifically, he showed that for each rational $t \geq 2$ and $\varepsilon > 0$, there is a graph G with $\mu_D(G) = t$ and $\rho(G) < \varepsilon$; and for rational $t > 1$ and $\varepsilon > 0$ there is a graph G with $\mu_D(G) = t$ and $\rho(G) > 1 - \varepsilon$.

Collinearity

Three nodes of a graph are said to be *collinear* if they can be labeled u, v, w so that

(9.7) $d(u, v) + d(v, w) = d(u, w).$

The *collinearity ratio* $cr(G)$ of a graph G is the proportion of collinear triples of nodes in G. Thus,

$$cr(G) = \frac{\text{number of collinear triples}}{\binom{p}{3}}.$$

The concept of collinear triples was introduced by Doyle and Graver [DG1]. Capobianco [C4] introduced the concept of the collinearity ratio. In Figure 9.14, we display several graphs along with their collinearity

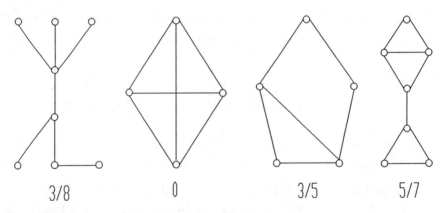

Figure 9.14 Graphs and their collinearity ratios.

ratios. Harary and Leep [HL1] described the concept of angles in graphs by applying trigonometric ideas to the sets of distances between nodes. They refer to the set of nodes satisfying (9.7) as *forming a straight angle* (0° or 180°). Harary and Melter [HM4] define an *equilateral triangle* in a graph G as a set $\{v_i, v_j, v_k\}$ of three nodes such that all three distances d_{ij}, d_{ik}, d_{jk} are finite and equal. They characterized the connected graphs having no equilateral triangles.

We note that inserting additional edges into a graph may increase, decrease, or not affect the value of $cr(G)$. What happens depends upon whether the number of new collinear triples formed exceeds the number of old ones which were destroyed when inserting the edges. Capobianco [C4] found the following relationship between the collinearity ratio and mean distance.

Theorem 9.16 For any connected graph G on $p \geq 3$ nodes, we have

(9.8) $cr(G) \geq 3(\mu_D(G) - 1)/(p - 2).$ □

A connected graph G is *geodetic* if any two nodes u, v are joined by precisely one path of length $d(u, v)$. These graphs were discussed in Chapter 7. We now present a result concerning the number of collinear triples in geodetic graphs. Letting $\tau(G)$ denote the number of noncollinear triples, we have

$$\tau(G) = \binom{p}{3} - \binom{p}{3} cr(G),$$

from which

$$cr(G) = 1 - \frac{\tau(G)}{\binom{p}{3}}.$$

On substituting this into (9.8), we obtain $\mu_D(G) \le (p+1)/3 - \tau(G)/\binom{p}{2}$, giving the following result of Doyle and Graver [DG1].

Theorem 9.17   If G is geodetic, then

$$\mu_D(G) = \frac{(p+1)}{3} - \frac{\tau(G)}{\binom{p}{2}}. \qquad \square$$

Doyle and Graver used Theorem 9.17 to establish the following tight bounds the mean distance of a nontrivial graph: $1 \le \mu_D(G) \le (p+1)/3$. They also extended their work to consider mean distance for digraphs [DG2]. In [C4], Capobianco discusses the relationship between collinearity and connectedness.

EXERCISES 9.3

1. Determine the distance distributions of K_p, P_p, and $W_{1,n}$.
2. The mean distance of C_p is

$$\mu_D(C_p) = \begin{cases} (p+1)/4 & \text{if } p \text{ is odd, and} \\ p^2/(4p-4) & \text{if } p \text{ is even.} \end{cases}$$

3. If $\binom{p}{2}$ is divisible by 3, then there are at least two graphs G on p nodes for which $\mu_D(G) = r(G)$, except for $\binom{p}{2} = 3, 6, 15,$ and 21.
4. Begin with the double star $S_{5,11}$ and construct a graph on 18 nodes with uniform distance distribution $(51,51,51)$.
5. For each d, except $d = 2$, there exists a graph G with diameter d for which $\mu_D(G) = r(G)$.
6. If G is self-complementary, then $\mu_D(G) \ge 3/2$ and this bound is best possible.   (Hendry[H21])
7. There is no graph G with a uniform distance distribution and even diameter for which $\mu_D(G) = r(G)$.
8. For line graphs, we have $dd(G) = dd(L(G))$ if and only if G is a cycle.   (Buckley [B22])
9. An r-regular graph G has $dd(G) = dd(\overline{G})$ if and only if $p = 4m+1$ with $m \ge 1$, $r = 2m$, and $d(G) = 2$.   (Buckley [B25])

10. A set X of nodes is a *dominating set* of G if every node in $V - X$ is adjacent to at least one node in X. Suppose that $d(G) = 3$. Then $dd(G) = dd(\overline{G})$ if and only if $D_1 = p(p-1)/4$ and D_3 is the number of pairs of adjacent nodes that form a dominating set of G.

11. The following conditions are equivalent:

    a. $dd(G) = dd(\overline{G})$ and $d(G) = 2$,

    b. $D_1 = p(p-1)/4$ and $d(G) = 2$,

    c. $D_1 = p(p-1)/4$ and no pair of adjacent nodes of G forms a dominating set.

12. Let T be a tree on p nodes with line graph $L(T)$. Then

$$\mu_D(L(T)) = p(\mu_D(T) - 1)/(p - 2).$$

13. If G has a uniform distance distribution and $\mu_D(G) = r(G)$, then the diameter d is odd and $r = (d - 1)/2$.

14. Determine the collinearity of each of the following graphs:
    a. K_p b. P_p c. C_p d. $K_{a,b}$ e. $W_{1,n}$.

9.4 PATH SEQUENCES

We now take up three sequences concerning the lengths of paths in graphs, called the path length distribution, the path degree sequence, and the geodesic distribution.

The Path Length Distribution

Chronologically, this was the second graphical sequence to be studied (the degree sequence was first). Let ℓ_i be the number of pairs of nodes joined by a path of length i. Capobianco [C3] defined the *path length distribution (pld)* of a connected graph G as the sequence $(\ell_1, \ell_2, \ldots, \ell_{p-1})$. Figure 9.15 illustrates this sequence.

Faudree, Rousseau, and Schelp [FRS1] showed that that there are no trees of order at most eight sharing a common *pld*. They proved, however, that for larger orders the situation is much different.

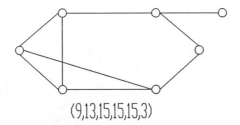

$$(9,13,15,15,15,3)$$

Figure 9.15   A graph and its path length distribution.

Theorem 9.18   For every $p \geq 9$ there are pairs of trees on p nodes with the same *pld*. Moreover, for any integer n, one can construct n trees having the same *pld*.   □

For any tree T, the sequence $pld(T)$ will end with a string of zeros beyond ℓ_d. There is a unique path between any pair of nodes in a tree. Thus for any tree T the first d terms of $pld(T)$ corresponds precisely to $dd(T)$. Thus we have the following.

Corollary 9.18   The distance distribution distinguishes nonisomorphic trees only for $p \leq 8$.   □

Faudree and Schelp [FS1,2] studied the *pld* in relation to the property of hamiltonian-connectedness. They conjectured that for any two nodes u and v in a hamiltonian-connected graph G, there is a path of length k joining u and v for all integers k, $p/2 \leq k \leq p - 1$. Using the dodecahedron and generalizations of it, Thomassen [T6] found infinitely many counterexamples. In particular, the dodecahedron has no cycles of length 19, so if $d(u, v) = 2$, then there is no path of length 18 joining u and v.

The Path Degree Sequence

For each node v in a connected graph G, let $p_i(v)$ be the number of paths of length i beginning at v. Then define the *path degree sequence of* v as

$$pds(v) = (p_0(v), p_1(v), p_2(v), \ldots, p_{p-1}(v)).$$

The sequences $pds(v)$ generally end with a string of zeros, so we terminate the sequence at the last nonzero term. Note the following:

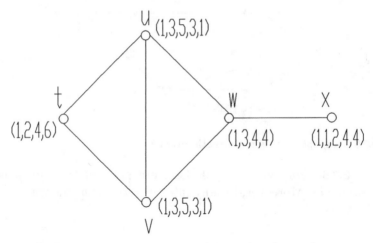

Figure 9.16   The path degree sequences of the nodes of a graph.

1. $p_0(v) = 1$ for all v.
2. $p_1(v) = \deg v$.
3.   If G is a tree, then $pds(v) = dds(v)$.

The *path degree sequence* $pds(G)$ of a graph G consists of the collection of sequences $pds(v)$ of its nodes, listed in numerical order. If a particular *pds* appears k times, we list it once with k as an exponent to indicate the multiplicity. For example, in Figure 9.16, $pds(t) = (1,2,4,6)$, $pds(w) = (1,3,4,4)$, and

$$pds(G) = ((1,1,2,4,4);\ (1,2,4,6);\ (1,3,4,4);\ (1,3,5,3,1)^2).$$

Since a tree T has a unique path joining each pair of nodes, clearly $pds(T) = dds(T)$. In general, $pds(G)$ distinguishes between nonisomorphic graphs far more frequently than $dds(G)$ does. In Figure 9.7, we showed the smallest pair of nonisomorphic graphs with the same *dds*, and these had order 5. Randic [R1] verified empirically that the smallest pair of graphs with identical *pds* must have order at least 12. The pair of trees in Figure 9.6 provide an example of nonisomorphic trees with identical *pds*. If we do not insist on trees, a slightly smaller such pair of graphs could be obtained by extending Slater's method [S20]. By using Slater's technique, in 1982 A.T. Balaban, G.S. Bloom, and L.V. Quintas constructed examples of pairs of nonisomorphic regular graphs with the same *pds*. Their rather large graphs thereby provided a partial answer to a question in [QS1]. It is still an open problem to find minimum order pairs of nonisomorphic k-regular graphs having the same *pds*.

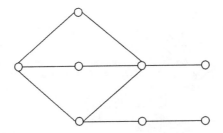

Figure 9.17   A graph with geodesic distribution $(22, 5, 1)$.

The Geodesic Distribution

Let g_i denotes the number of pairs of nodes joined by i geodesics (shortest paths) in graph G. Capobianco defined the *geodesic distribution* $gd(G)$ of a graph G as $(g_1, g_2, g_3, \ldots)$. Recall from §7.3 that a *geodetic graph* has a unique geodesic joining each pair of nodes. Thus, G is geodetic if and only if $gd(G)$ has a single term. A graph and its geodesic distribution is given in Figure 9.17.

The first difference we note between $gd(G)$ and other sequences is that the length of the sequence $gd(G)$ is not specified. This length varies with G and can be quite long. Let $m(p)$ denote the maximum length of $gd(G)$ for a graph on p nodes. M. Capobianco described the following to us.

Theorem 9.19   Let $\mathcal{T} = \{\langle t_i \rangle\}$ be the set of all partitions of the integer $p - 2$. The value $m(p)$ is achieved by maximizing $\prod t_i$ over $\mathcal{T}$.   □

Buckley [B25] determined the exact value of $m(p)$ with the following results.

Lemma 9.20a   In $m(p) = \prod t_i$, each factor t_i is at most 4.   □

Lemma 9.20b   For $m(p) = \prod 2^b 3^c$, where $2b + 3c = p - 2$, we have $b \leq 2$.   □

Theorem 9.20   The maximum number of geodesics between a pair of nodes in a graph of order p is given by

$$m(p) = \begin{cases} 2^2 \cdot 3^{(p-6)/3} & p \equiv 0 \pmod 3, \\ 2 \cdot 3^{(p-4)/3} & p \equiv 1 \pmod 3, \\ 3^{(p-2)/3} & p \equiv 2 \pmod 3. \end{cases}$$ □

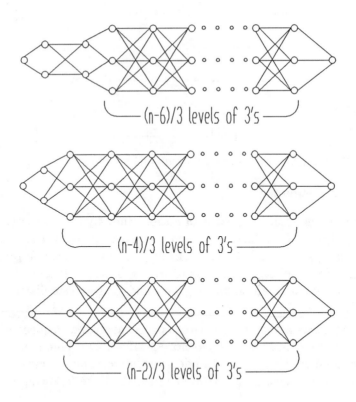

Figure 9.18   Graphs on p nodes with the maximum possible number of geodesics between a given pair of nodes.

The maximum values in Theorem 9.20 are achievable. Graphs realizing the maximum are displayed in Figure 9.18.

Analogous to the mean distance, we can define the *geodesic mean* of a graph on p nodes as

$$\mu_g(G) = \sum_{i=1}^{m(p)} i \cdot g_i \bigg/ \binom{p}{2}.$$

Graph G is geodetic if and only if $\mu_g(G) = 1$. One can use $\mu_g(G)$ to measure how close G is to being geodetic. The larger $\mu_g(G)$ is, the farther G is from being geodetic. Buckley [B26] defined a graph G on p nodes to be *antigeodetic* if $\mu_g(G) \geq \mu_g(H)$ for all graphs H on p nodes and proposed the open problem of characterizing such graphs.

EXERCISES 9.4

1. Construct two trees of order 9 having the same path length
    distributions.

2. Determine $pds(G)$ for each of the following:
    a. K_p b. P_p c. C_p d. $K_{a,b}$ e. $W_{1,n}$.

3. The sequence $dds(G) = pds(G)$ if and only if G is a tree.

4. When p is odd, the geodesic distribution of C_p has only one term,
    $\binom{p}{2}$. If p is even, then $gd(C_p) = (p(p-2)/2, p/2)$.

5. The geodesic distribution of $W_{1,n}$ is

$$gd(W_{1,n}) = \begin{cases} (6) & \text{for } n = 4. \\ (8, 0, 2) & \text{for } n = 5. \\ \left(\binom{n-1}{2}, n-1\right) & \text{for } n \geq 6. \end{cases}$$

9.5   OTHER SEQUENCES

There are several other sequences that are distance related which have
not yet been discussed.

Common Neighbor Distribution

Let n_i denote the number of pairs of nodes with i common neighbors in a
graph G. The *common neighbor distribution* $nd(G)$ of a graph on p nodes
was defined by Buckley [B24] as $(n_0, n_1, n_2, \ldots, n_{p-2})$. This sequence was
introduced to aid in distinguishing nonisomorphic graphs. For trees T,
$nd(T)$ is derivable from $dd(T)$.

Theorem 9.21   For a tree T, $nd(T) = (\binom{p}{2} - D_2, D_2)$.   □

Since D_2 is derivable from the degree sequence for a tree, we have the
following.

Corollary 9.21   Let $\deg v_i$ denote the degree of node v_i in a tree T.
Then

$$nd(T) = \left(\binom{p}{2} - \sum_{i=1}^{p} \binom{\deg v_i}{2}, \sum_{i=1}^{p} \binom{\deg v_i}{2}\right).$$ □

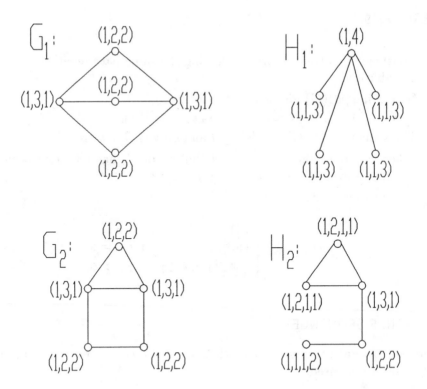

Figure 9.19   Pairs of graphs for which exactly one of $dd(G)$ and $nd(G)$ agree.

Since $nd(G)$ is derivable from $dd(G)$ for trees, it is only worth examining $nd(G)$ for nontree graphs. It is interesting to note that there are pairs of graphs for which $dd(G)$ agree but $nd(G)$ differ as well as pairs for which $nd(G)$ agree but $dd(G)$ differ. We illustrate this in Figure 9.19, where $dd(G_1) = dd(G_2) = (6,4)$ while $nd(G_1) = (6,0,3,1)$ and $nd(G_2) = (3,5,2)$. For graphs H_1 and H_2, $nd(H_1) = nd(H_2) = (4,6)$ while $dd(H_1) = (4,6)$ and $dd(H_2) = (5,3,2)$.

Note that graphs G_1 and G_2 of Figure 9.19 are the smallest pair of nonisomorphic graphs having the same distance degree sequence. Thus, it is sometimes possible to distinguish nonisomorphic graphs using $nd(G)$ when $dds(G)$ cannot distinguish them. A smallest pair of connected nontree graphs for which both $nd(G)$ and $dds(G)$ agree is shown in Figure 9.20. We note that although neither $nd(G)$ nor $dds(G)$ can distinguish these graphs, $gd(G)$ can.

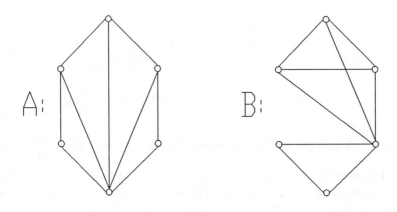

$$dds(A) = dds(B) = ((1,2,3),\ (1,3,2),\ (1,5))$$
$$nd(A) = nd(B) = (9,6)\ \text{but}\ \ gd(A) = (12,3) \neq gd(B) = (15)$$

Figure 9.20   Graphs A and B with $dds(A) = dds(B)$ and $nd(A) = nd(B)$.

Just as we defined arithmetic means of other sequences, we define the *mean number of common neighbors* as

(9.9)
$$\mu_N(G) = \sum_{i=0}^{p-2} i\, n_i \Big/ \binom{p}{2}.$$

Buckley [B24] showed that $\mu_N(G)$ is easily derived from the degree sequence of G.

Theorem 9.22   In any graph G,

(9.10)
$$\mu_N(G) = \sum_{i=0}^{p-2} \binom{\deg v_i}{2} \Big/ \binom{p}{2}. \qquad\qquad \square$$

It was found in [B25] that when the common neighbor distribution of a graph equals that of its complement, there is a direct relation to dominating sets. Set $X \subset V(G)$ *dominates* set $Y \subset V(G)$ if every node in $Y - X$ is adjacent to a node in X. Recall that X is a *dominating set* of G if X dominates $V(G)$.

Theorem 9.23   If $nd(G) = nd(\overline{G})$, then n_0 equals the number of dominating sets of order 2 in G.

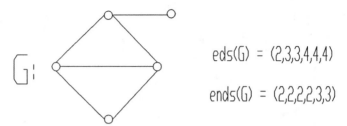

Figure 9.21   A graph G with its sequences $eds(G)$ and $ends(G)$.

Proof   Let $nd(G) = nd(\overline{G})$. Then for two distinct nodes u, v having no common neighbor, there exists a pair of nodes u', v' with no common neighbors in $\overline{G}$. Hence $\{u', v'\}$ is a dominating set for G.   □

Corollary 9.23   If G is a connected graph with $nd(G) = nd(\overline{G})$ and $n_0 > 0$, then $d(G) \leq 5$.   □

The following result was also given in [B25].

Theorem 9.24   For a graph G, $nd(G) = nd(\overline{G})$ if and only if n_i equals the number of pairs of nodes which dominate $n - i$ nodes of G.   □

Edge Degree Sequences

Let $ed(x)$ be the number of edges incident with edge x. Then the *edge degree sequence* $eds(G)$ of a graph G is the sequence $ed_1, ed_2, \ldots, ed_q$ of values $ed(x)$, arranged in nondecreasing order. Note that if $x = uv$, then $ed(x) = \deg u + \deg v - 2$. Clearly, $eds(G)$ is precisely the degree sequence of $L(G)$, the line graph of G. Realizability questions for $eds(G)$ were studied by Bauer [B2]. He characterized the realizable sequences in the cases when $\max\{ed_i\} = q - 1$ or $q - 2$; $\max\{ed_i\} \leq 3$; and $\max\{ed_i\} = \min\{ed_i\}$, i.e, $L(G)$ is regular.

For each edge $x = uv$, let the *edge-to-node degree* $end(x)$ of x be the number of distinct nodes in $G - x$ adjacent to either u or v. Then the *edge-to-node degree sequence* $ends(G)$ of G is the sequence of values $end(x)$ listed in nondecreasing order. This concept was introduced and studied by Bauer, Bloom and Boesch [BBB1]. Among their results, they characterized regular (that is, having all $end(x)$ equal) sequences $ends(G)$ realizable by triangle-free graphs. A graph G along with $eds(G)$ and $ends(G)$ is displayed in Figure 9.21.

EXERCISES 9.5

1. Determine $nd(G)$ for each of the following:
   a. K_p b. P_p c. C_p.

2. Let G be connected with $p \geq 3$. Then $n_0 = 0$ if and only if $d(G) \leq 2$ and every edge of G is contained in a triangle.

3. The common neighbor distribution of the wheel $W_{1,n}$ is

$$nd(W_{1,n}) = \begin{cases} (0,4,4,2) & n = 4; \\ (0,(n-1)(n-4)/2, 2(n-1)) & n \neq 4. \end{cases}$$

4. The common neighbor distribution of $K_{a,b}$ is
   $nd(K_{a,b}) = (ab, n_1, n_2, \ldots, n_{p-2})$, where $n_a = \binom{b}{2}$, $n_b = \binom{a}{2}$ and $n_i = 0$ for all other $i \geq 1$ when $2 \leq a < b$; and
   $n_a = \binom{a}{2} + \binom{b}{2}$ and $n_i = 0$ for all other $i \geq 1$ when $2 \leq a = b$.

5. Graph G is triangleless if and only if $n_0 = \binom{p}{2} - D_2$.

6. Determine $eds(G)$ and $ends(G)$ for each of the following:
   a. K_p b. P_p c. C_p d. $W_{1,n}$.

7. A graph G has diameter 2 if and only if
   $\max\{ed(x) : x \in E(G)\} \leq p - 3$.
                                        (G.S. Bloom, J.W. Kennedy, and L.V. Quintas)

8. Use the equivalence of equations (9.9) and (9.10) to prove the combinatorial identity

$$\sum_{i=1}^{p-1}(p-1)\binom{p}{i} = p \cdot 2^{(p-1)}.$$

FURTHER RESULTS

Buckley and Superville [BS7] determined the mean distance for various classes of graphs. We now state several of their results.

1. The mean distance of $K_{a,b}$ is

$$\mu_D(K_{a,b}) = \frac{2(a^2 - a + b^2 - b + ab)}{(a^2 - a + b^2 - b + 2ab)}.$$

2.   For nontrivial paths and for cycles, we have

  a. $\mu_D(P_p) = r(P_p)$ if and only if $p = 2$ or 5 $(p \geq 2)$.

  b. $\mu_D(C_p) = r(C_p)$ if and only if $p = 3$.

3.   For complete bipartite graphs and wheels, we have

  a. $\mu_D(K_{a,b}) = r(K_{a,b})$ if and only if $a = b = 1$;

  b. $\mu_D(W_{1,n}) = r(W_{1,n})$ if and only if $n = 4$.

The *independence number* of a graph G is the maximum number of nodes in G, no two of which are adjacent. The mean distance of a connected graph is less than or equal to the the independence number of G.   (Chung [C11])

An f-*tree* is a tree whose maximum degree is at most f. The 4-trees are of special significance as they contain the set of trees that represent saturated hydrocarbons (alkanes) important in organic chemistry. In various applications, such as crystal formation, a molecule is assumed to be embedded in some space, generally a lattice. Kennedy and Quintas [KQ1] studied embedding problems for molecules, especially ones modeled by f-trees. For trees T they define a sequence called the *maximum reach sequence* for T relative to lattice $\mathcal{L}$, derivable from the distance degree sequence for T. They use this sequence to establish necessary conditions for the embeddability of T in $\mathcal{L}$.

Bloom, Kennedy, and Quintas [BKQ2] pose a number of problems concerning DDI graphs. They asked whether there exists a graph G such that both G and $\overline{G}$ are DDI. They also ask for the smallest order (or diameter) for a k-regular DDI graph.

A graph G is *bigeodetic* if each pair of nodes are joined by at most two geodesics. Alagar, Opatrny, and Srinivasan [AOS1] characterize bigeodetic graphs, develop constructions, and show that for a (p,q)-bigeodetic graphs of diameter d, $p - 1 \leq q \leq \binom{p-d+1}{2} + d$.

CHAPTER 10

Digraphs

There is so much to digraph theory that several books have been written on the subject [HNC1], [FR1], and [M11], the latter presenting the subject with tournaments. Throughout this chapter we shall emphasize those properties of digraphs which set them apart from graphs, giving particular attention to distance related properties. We close the chapter with a brief discussion of tournaments.

10.1 DIGRAPHS AND CONNECTEDNESS

We have already seen all the digraphs with 3 nodes and 3 arcs in Fig. 1.3. For completeness, we begin with a few definitions, including a few from Chapter 1. A *digraph D* consists of a finite set V of nodes and a collection of ordered pairs of distinct nodes from V. Any such pair (u,v) is called an *arc* or *directed edge* and will be denoted uv. The arc uv goes *from u to v* and is *incident with* u and v. We also say that u is *adjacent to* v and v is *adjacent from* u. The *indegree* $id(v)$ of a node v is the number of nodes adjacent to v, and the *outdegree* $od(v)$ is the number adjacent from v.

A *(directed) walk* in a digraph D is an alternating sequence of nodes and arcs $v_0, x_1, v_1, \ldots, x_n, v_n$ in which each arc x_i is $v_{i-1}v_i$. The *length* of such a walk is n, the number of arcs in it. A *closed walk* has the same first and last nodes, and a *spanning walk* contains all the nodes of D. A *path* is a walk in which all nodes are distinct; a *cycle* is a nontrivial closed walk with all nodes distinct (except the first and last). An *acyclic digraph* contains no directed cycles. If there is a path from u to v, then v is said to be *reachable from* u, and the *distance* $d(u, v)$ from u to v is the length of any shortest such path.

Each walk is directed from the first node v_0 to the last v_n. We also need a concept which does not have this property of direction and is analogous to a walk in a graph. A *semiwalk* is again an alternating sequence $v_0, x_1, v_1, \ldots, x_n, v_n$ of nodes and arcs, but each arc x_i may be either $v_{i-1}v_i$ or $v_i v_{i-1}$. A *semipath*, *semicycle*, and so forth, are defined as expected.

Whereas a graph is either connected or is not, there are three different ways in which a digraph may be connected. A digraph is *strongly connected*, or *strong*, if every two nodes are mutually reachable; it is *unilaterally connected*, or *unilateral*, if for any two nodes at least one is reachable from the other; and it is *weakly connected*, or *weak* if every two nodes are joined by a semipath. Clearly, every strong digraph is unilateral and every unilateral digraph is weak, but the converse statements are not true. A digraph is *disconnected* if it is not even weak. We note that the *trivial digraph*, consisting of exactly one node, is vacuously strong since it does not contain two nodes.

We may now state necessary and sufficient conditions for a digraph to satisfy each of the three kinds of connectedness.

Theorem 10.1   A digraph is strong if and only if it has a closed spanning walk, it is unilateral if and only if it has a spanning walk, and it is weak if and only if it has a spanning semiwalk.   □

Corresponding to connected components of a graph, there are three different kinds of components of a digraph D. A *strong component* of D is a maximal strong subgraph; a *unilateral component* and a *weak component* are defined similarly. It is very easy to verify that every node of a digraph D is in just one weak component and in at least one unilateral component, and this also holds for each arc. Furthermore, each node is in exactly one strong component, and an arc lies in one strong component or none, depending on whether or not it is in some cycle.

The strong components of a digraph are the most important among these. One reason is the way in which they yield a new digraph which, although simpler, retains some of the structural properties of the original. Let $S_1, S_2, \ldots, S_n$ be the strong components of D. The *condensation* D^* (illustrated in Figure 10.1) of D has the strong components of D as its

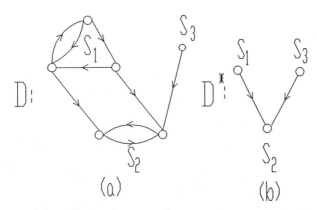

Figure 10.1   A digraph and its condensation.

nodes, with an arc from S_i to S_j whenever there is at least one arc in D from a node of S_i to a node of S_j.

It follows from the maximality of strong components that the condensation D^* of any digraph D is acyclic. Obviously, the condensation of any strong digraph is the trivial digraph.

Orientations and Strong Digraphs

A digraph D is *symmetric* if whenever uv is an arc, then so is vu. On the other hand, D is *asymmetric* if the presence of uv obviates that of vu. When both uv and vu are in D, they form a *symmetric pair*. This D is symmetric if and only if it has no symmetric pairs. Figure 10.2 illustrates digraphs with these properties. The *digraph of a graph* $G = (V, E)$, written $\mathbf{D}(G)$, also has V as its node set, and each edge e of G is replaced by the symmetric pair of arcs joining the two endnodes of e. The *(underlying) graph of a digraph* D, written $\mathbf{G}(D)$, also has the same node set as D, but two nodes u and v are now adjacent if they are joined in D by at least one arc. Obviously $\mathbf{G}(\mathbf{D}(G)) = G$ but $\mathbf{D}(\mathbf{G}(D))$ need not be D; it is the "symmetric closure" of D. These two operations will be found useful.

An *orientation of a graph* G is any digraph that results from an assignment of directions to the edges of G. If G has at least one edge, any orientation of G is asymmetric and is called an *oriented graph*. Robbins [R7] characterized those graphs that have a strongly connected orientation.

Theorem 10.2   A graph G has a orientation that is strong if and only if G is connected and has no bridges.   □

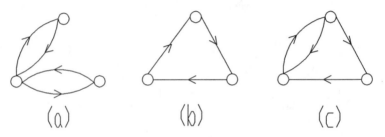

Figure 10.2   Three strong digraphs - one symmetric, one asymmetric, and one neither.

Since every nontrivial graph G has at least two nodes that are not cutnodes, it follows that every nontrivial digraph does as well. This means that if D is strong, unilateral, or weak, then there are two nodes u and v such that both $D - u$ and $D - v$ are weak.

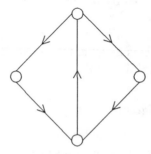

Figure 10.3   A strong digraph with no spanning cycle.

It should be stressed that a strong digraph need not have a spanning cycle. For example, the digraph on 4 nodes in Figure 10.3 is strong yet has no spanning cycle. However, there is a strong (no pun intended) relationship between cycles and strong digraphs as illustrated in the following characterization.

Theorem 10.3   A weak digraph D is strong if and only if every arc of D is contained in a cycle.

Proof   If D is strong then for every arc uv there must be a path $v, e_0, v_1, e_1, \ldots, e_n, u$ from v to u. Then $v, e_0, v_1, e_1, \ldots, e_n, u, uv, v$ is a cycle containing arc uv.

Conversely, it is given that every arc of the weak digraph D is contained in a cycle. Since D is weak, there is a semiwalk $u, e_1, v_1, e_2, \ldots, e_n, v$ joining any two nodes u and v. Arc e_1 (which could be either uv_1 or v_1u)

Table 10.1   SIZES FOR CONNECTEDNESS CATEGORIES

Category	Minimum Number of Arcs	Maximum Number of Arcs
0	0	$(p-1)(p-2)$
1	$p-1$	$(p-1)(p-2)$
2	$p-1$	$(p-1)^2$
3	p	$p(p-1)$

is contained in a cycle, so u and v_1 are in the same strong component of D. Similarly each v_i and v_{i+1} are in the same strong component as are v_n and v. Thus all the nodes of the semiwalk joining u and v are in the same strong component of D. As u and v are any two nodes, it follows that D is strong.   □

EXERCISES 10.1

1. A digraph is *strictly weak* if it is weak but not unilateral; it is *strictly unilateral* if it is unilateral but not strong. Let C_0 contain all disconnected digraphs, C_1 the strictly weak ones, C_2 strictly unilateral, and C_3 those which are strong. Then the maximum and minimum number q of arcs among all p node digraphs in connectedness category C_i, $i = 0$ to 3 is given in Table 10.1.
   (Cartwright and Harary [CH1])

2. A digraph is unilateral if and only if its condensation has a unique spanning path.

3. Determine all nonisomorphic digraphs having 4 nodes and 4 arcs.

4. The *cartesian product* $D_1 \times D_2$ of two digraphs has $V_1 \times V_2$ as its node set, and (u_1, u_2) is adjacent to (v_1, v_2) whenever $[u_1 = v_1$ and u_2 adj $v_2]$ or $[u_2 = v_2$ and u_1 adj $v_1]$. (Note that this is defined just as for graphs except that adjacency is directed.) When D is in the connectedness category C_n, we write $c(D) = n$ for $n = 0, 1, 2, 3$. Then $c(D_1 \times D_2) = \min\{c(D_1), c(D_2)\}$ unless $c(D_1) = c(D_2) = 2$ in which case $c(D_1 \times D_2) = 1$.   (Harary and Trauth [HT1])

5. No strictly weak digraph contains a node whose removal results in a strong digraph.   (Harary and Ross [HR3])

6. A digraph is *r-regular* if $id(v) = od(v) = r$ for each node v of D. Determine conditions on p and r which guarantee the existence of an r-regular digraph D on p nodes.

7. Let D be a digraph with p nodes and q arcs with $V(D) = \{v_1, v_2, \ldots, v_p\}$. Then

$$\sum_{i=1}^{p} \mathrm{od}(v_i) = \sum_{i=1}^{p} \mathrm{id}(v_i) = q.$$

8. The *line digraph* $L(D)$ has as its nodes the arcs of the given digraph D and x is adjacent to y in $L(D)$ whenever arcs x, y induce a walk in D. Calculate the number of nodes and arcs of $L(D)$ in terms of D.   (Harary and Norman [HN4])

10.2   ACYCLIC DIGRAPHS

The *converse digraph* D' of D has the same set of nodes as D and the arc uv is in D' if and only if arc vu is in D. Thus the converse of D is obtained by reversing the direction of every arc of D. We have already encountered some converse concepts, such as indegree and outdegree, and these concepts concerned with direction are related by a rather powerful principle. This is a classical result in the theory of binary relations.

Principle of Directional Duality   For each theorem about digraphs, there is a corresponding theorem obtained by replacing every concept by its converse.

We now illustrate how this principle generates new results.

Theorem 10.4   An acyclic digraph has at least one node of outdegree zero.

Proof   Consider the last node of any longest path in the digraph. This node can have no nodes adjacent from it since otherwise there would be a cycle.   □

The dual theorem follows immediately by applying the Principle of Directional Duality. In keeping with the use of D' to denote the converse of digraph D, we shall use primes to denote dual results.

Theorem 10.4′   An acyclic digraph D has at least one node of indegree zero.   □

It was noted that the condensation of any digraph is acyclic. The it adjacency matrix $A(D)$ of a digraph D is a $(0,1)$-matrix with $a_{ij} = 1$ if there is an arc from v_i to v_j. We now provide several characterizations.

Theorem 10.5   The following properties of a digraph D are equivalent.

1. D is acyclic.
2. D^* is isomorphic to D.
3. Every directed walk of D is a directed path.
4. It is possible to label the nodes of D so that the adjacency matrix $A(D)$ is upper triangular. □

Out-Trees

Two dual types of acyclic digraphs are of particular interest. A *source* in D is a node which can reach all others; a sink is the dual concept. An *out-tree* is a digraph with a source but having no semicycles; an *in-tree* is its dual, see Figure 10.4. The source of an out-tree is its *root* as is the sink of an in-tree.. An out-tree has also been called an *arborescence*. These concepts have been widely used by computer scientists in searching and sorting algorithms.

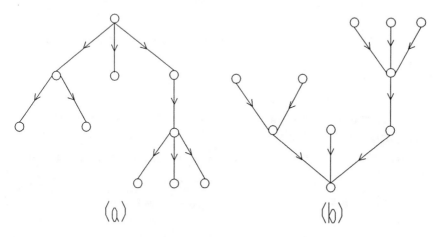

Figure 10.4   An out-tree and its converse in-tree.

Theorem 10.6   A weak digraph is an out-tree if and only if it has exactly one root and all other nodes have indegree one.

Proof   Suppose that D is a weak digraph. If D is an out-tree, it has exactly one root and no semicycles. Hence each node is reachable from the root in only one way, so each nonroot has indegree one. On the other hand, if D has exactly one root r and all other nodes have indegree one, then there is a unique directed path from r to each other node, and D has no semicycles. Thus D is an out-tree.   □

Theorem 10.6′   A weak digraph is an in-tree if and only if it has exactly one root and all other nodes have outdegree 1.   □

EXERCISES 10.2

1. A digraph D is an out-tree if and only if D contains a node r such that there is precisely one directed path from r to each other node of D.

2. Call a digraph *resourceful* if every two nodes can be reached from a common node. Then D is resourceful if and only if it has a source.

3. Describe the dual concept to that of a digraph being resourceful.

4. If every node of D has positive outdegree, then D contains a directed cycle.

5. A digraph D is acyclic if and only if it is possible to label its nodes $v_1, v_2, \ldots, v_p$ so that any arc $v_i v_j$ of D satisfies $i < j$.

6. In any labeling of the nodes of D as in Exercise 5, if v_j is reachable from v_i then $i < j$.

7. Give an example of a digraph which is not a directed cycle that is isomorphic with its converse.

8. Characterize matrices which are adjacency matrices of digraphs.

9. Every walk in an acyclic digraph is a path.

10. The following statements are equivalent for a digraph D:

     a. D is an out-tree.

     b. D is resourceful and has $p - 1$ arcs.

     c. D is resourceful and has no semicycles.

10.3   MATRICES AND EULERIAN DIGRAPHS

There are several matrices associated with a digraph, and each one provides certain information about the digraph it represents. For example, the row sums of the adjacency matrix $A(D)$ of a digraph D give the outdegrees of the nodes of D, while the column sums give the indegrees.

As in the case of graphs, the powers of the adjacency matrix A of a digraph give information about the number of walks from one node to another.

Theorem 10.7   The i, j entry $a_{ij}^{(n)}$ of A^n is the number of walks of length n from v_i to v_j.   □

Three other matrices associated with D are the reachability matrix, the distance matrix, and the detour matrix. In the *reachability matrix* $R(D)$, $r_{ij} = 1$ if v_j is reachable from v_i and 0 otherwise. The i, j entry of the *distance matrix* $\partial(D)$ gives the distance from node v_i to node v_j, and is ∞ if there is no path from v_i to v_j. These two matrices were introduced by Harary [H14]. In the *detour matrix* $T(D)$, the i, j entry is the length of any longest path from v_i to v_j, and again is ∞ if there is no such path. These three matrices for the digraph of Figure 10.5 are

$$
\begin{pmatrix}
1 & 0 & 0 & 0 & 0 \\
1 & 1 & 1 & 1 & 0 \\
1 & 0 & 1 & 0 & 0 \\
1 & 0 & 1 & 1 & 0 \\
0 & 0 & 0 & 0 & 1
\end{pmatrix}
\quad
\begin{pmatrix}
0 & \infty & \infty & \infty & \infty \\
1 & 0 & 1 & 1 & \infty \\
1 & \infty & 0 & \infty & \infty \\
2 & \infty & 1 & 0 & \infty \\
\infty & \infty & \infty & \infty & 0
\end{pmatrix}
\quad
\begin{pmatrix}
0 & \infty & \infty & \infty & \infty \\
3 & 0 & 2 & 1 & \infty \\
1 & \infty & 0 & \infty & \infty \\
2 & \infty & 1 & 0 & \infty \\
\infty & \infty & \infty & \infty & 0
\end{pmatrix}
$$

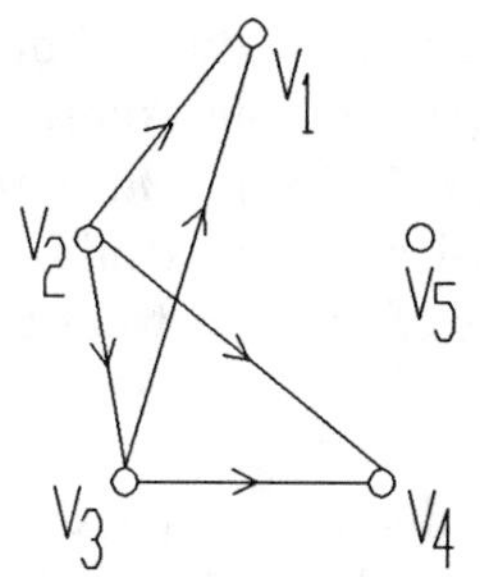

$$
R(D) \qquad\qquad\qquad \partial(D) \qquad\qquad\qquad\qquad T(D)
$$

Figure 10.5   A digraph to illustrate three associated matrices.

Corollary 10.7a   The entries of the reachability and distance matrices can be obtained from the powers of A as follows:

1.   For all i, $r_{ii} = 1$ and $d_{ii} = 0$.

2. $r_{ij} = 1$ if and only if for some n, $a_{ij}^{(n)} > 0$.

3. $d(v_i, v_j)$ is the least n (if any) such that $a_{ij}^{(n)} > 0$, and 0 otherwise. $\square$

Unlike the situation for the reachability and distance matrices, there is no efficient method for finding the detour matrix. In fact, the problem of finding the detour matrix is $\mathcal{NP}$-complete.

The *elementwise product* (sometimes called the *Hadamard product*) $B \times C$ of two matrices $B = [b_{ij}]$ and $C = [c_{ij}]$ has $b_{ij}c_{ij}$ as its i, j entry. The reachability matrix can be used with elementwise products to find strong components [H14].

Corollary 10.7b   Let v_i be a node of a digraph D. The strong component of D containing v_i is determined by the unit entries in the ith row of the symmetric matrix $R \times R^T$. $\square$

In Chapter 7, we discussed distance matrix realizability problems of graphs. The corresponding problem for digraphs was studied by Simões-Pereira [S11]. As in the case for graphs, weights are placed on the arcs corresponding to a (directed) distance from one node to another. If for some pair of nodes v_i and v_j, there is no intermediate node v_k such that $d_{ij} = d_{ik} + d_{kj}$, then d_{ij} is called a basic distance. Thus each basic distance d_{ij} is determined uniquely by the weight on the single arc $v_i v_j$. In Figure 10.6, $d_{13} = 4$ and is basic, but d_{24} is not basic because $d_{24} = d_{23} + d_{34}$.

A weighted digraph W is a realization of matrix M of order n if there is a subset $V = \{v_1, v_2, \ldots, v_n\}$ of the nodes of W having order n such that $d(v_i, v_j) = m_{ij}$ for all i, j, $1 \le i, j \le n$. A realization is optimal if the total of the weights used on its arcs is a minimum. Optimal realizations of directed distance matrices of order n were characterized in [S11].

Theorem 10.8   A directed distance matrix M of order n has an optimal realization if and only if M can be realized by a simple directed cycle, or equivalently, M has n basic entries. $\square$

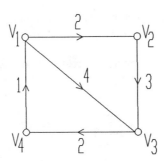

Figure 10.6   A weighted digraph.

Eulerian Digraphs

The number of spanning in-trees to a given node in a digraph was found by Bott and Mayberry [BM2] and proved by Tutte [T13a]. To give this result, called the matrix tree theorem for digraphs, we need some other matrices related to D. Let M_{od} denote the matrix obtained from $-A$ by replacing the ith diagonal entry by $od(v_i)$. The matrix M_{id} is defined dually.

Theorem 10.9   For any labeled digraph D, the value of the cofactor of each entry in the ith row of M_{od} is the number of spanning in-trees with v_i as sink.   □

Theorem 10.9′   The value of the cofactor of any entry in the jth column of M_{id} is the number of spanning out-trees with root v_i.   □

An *eulerian trail* in a digraph D is a closed spanning walk in which each arc of D occurs exactly once. A digraph is *eulerian* if it has such a walk.

Theorem 10.10   For any weak digraph D, the following statements are equivalent:

1. D is eulerian.
2. For each node u, $od(u) = id(u)$.
3. There exists a partition of the arc set of D into directed cycles.   □

We will now state a theorem giving the number of eulerian trails in an eulerian digraph. This result was discovered independently by de Bruijn and van Aardenne-Ehrenfest [BE1] and Smith and Tutte [ST1]. It can be elegantly proved using the matrix tree theorem for digraphs, as in Kasteleyn [K1,p.76]; see also Harary and Palmer [HP3,p.28].

Corollary 10.9   In an eulerian digraph D, let $d_i = id(v_i)$ and c be the common value of all the cofactors of M_{od}. Then the number of eulerian trails is

$$c \cdot \prod_{i=1}^{p} (d_i - 1)! \qquad \qquad \square$$

Note that for an eulerian digraph D, we have $M_{od} = M_{id}$ and all row sums as well as all column sums are zero, so that all cofactors are equal. For the digraph in Figure 10.7, $c = 7$ and there are 14 eulerian trails. Two of them are $v_1 v_2 v_3 v_4 v_2 v_1 v_3 v_1 v_4 v_1$ and $v_1 v_2 v_1 v_4 v_2 v_3 v_4 v_1 v_3 v_1$.

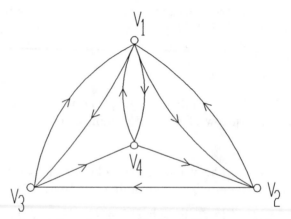

Figure 10.7   An eulerian digraph whose eulerian trails are counted.

EXERCISES 10.3

1. Characterize matrices which are reachability matrices of some digraph.

2. For the adjacency matrix A of an acyclic digraph $A^p = 0$.

3. Let D be a nontrivial weak digraph. Then D has an open spanning trail if and only if either D is eulerian or D contains nodes u and v such that $id(u) = od(u) + 1$ and $od(v) = id(v) + 1$ while all other nodes w satisfy $id(w) = od(w)$.

4. The number of eulerian trails of a digraph D equals the number of hamiltonian cycles of $L(D)$.   (Kasteleyn [K1])

10.4   LONG PATHS IN DIGRAPHS

The values of certain invariants might guarantee the existence of a path of a given length in a graph. A simplar statement is true for digraphs. For example, Roy [R8] and Gallai [G2] showed independently that every orientation of an n-chromatic graph contains a directed path of length $n-1$. In Chapter 4, we discussed hamiltonian graphs. We saw that many of the sufficient conditions for hamiltonicity have counterparts when considering generalizations of hamiltonian graphs. We shall now examine these concepts in the context of digraphs.

Diameter

For digraphs, distance concepts are defined analogous to those for graphs except that we must heed the directions on the arcs. Thus, the distance from u to v is the length of a shortest (directed) u-v path. The *eccentricity* of a node v in a digraph D is its distance to a farthest node in D. For a strong digraph the eccentricities are all finite. The *radius* is the minimum eccentricity and the *diameter* is the maximum. It is easy to read off the eccentricity of node $v_i \in D$ from the distance matrix $\partial(D)$. Simply find the largest entry in row i of $\partial(D)$. The diameter and radius of D can be obtained similarly. More efficient methods are given in Chapter 11.

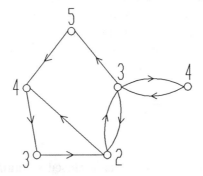

Figure 10.8   A digraph and its eccentricities.

Note that for digraphs, some familiar distance relations do not hold. For example, graphs have the property that the eccentricities of adjacent nodes differ by at most one, but this is not true even for strong digraphs. Also, the familiar relationship $r \leq d \leq 2r$ does not hold for digraphs. In Figure 10.8, we display a digraph D with radius $r(D) = 2$ and diameter $d(D) = 5$.

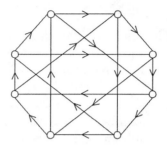

Figure 10.9   A directed circulant.

We recall for a moment a concept from §3.4. A *circulant* is a graph determined by its order p and a subset $S = \{a, b, c, \ldots\}$ of $\{1, 2, 3, \ldots, \lfloor p/2 \rfloor\}$ as follows. The *circulant* graph $C(p : S) = C(p : a, b, c, \ldots)$ has node set $Z_p = \{0, 1, \ldots, p - 1\}$ and each node u is adjacent with the nodes $u + a$, $u + b$, $u + c$, $\ldots$, all sums taken modulo p. Now the corresponding *directed circulant* is defined similarly, except that adjacent with is replaced by adjacent to. Certain directed circulants have also been called "double loop computer networks", namely, those of the form $C(p; 1, h)$ which is denoted more briefly by $D(p, h)$. To illustrate, $D(8, 3)$ is shown in Figure 10.9.

The exact value of the diameter of the directed circulant $D(p, h)$ is given by a complex formula due to Y. Cheng. One wants to minimize the diameter as it varies directly with the transmission time when the nodes are microprocessors and the arcs are communication channels. The choice of the jump size h is crucial in such problems. Hwang and Xu [HX1] obtained good asymptotic results on choosing h to minimize the diameter in directed circulants. Also see Cheng and Hwang [CH4] for results on the diameter of weighted circulants.

Hamiltonian Digraphs

A digraph is *hamiltonian* if it has a closed spanning path. The sufficient conditions for a digraph to be hamiltonian are similar in flavor to those for graphs. Of course, every hamiltonian digraph is strong, but the converse is not true. For example, a strong digraph with a cutnode is not hamiltonian. Thus, most results on hamiltonian digraphs begin with the assumption that D is strong. Several of the early results, beginning with that of Ghouila-Houri [G4], were subsumed by a result of Meyniel. Thus we first state Meyniel's result [M6], for which a simple proof was found by Bondy and Thomassen [BT4]. For a node v in a digraph D, let $\deg v = id(v) + od(v)$.

Theorem 10.11   If D is a nontrivial strong digraph of order p such that for every pair of distinct nonadjacent nodes u and v

(10.1) $$\deg u + \deg v \geq 2p - 1,$$

then D is hamiltonian.

Outline of Proof   Since D is strong it has cycles. Let C be a longest cycle in D and suppose that C does not span D. Then for a node v not on C there are paths from C to v and back to C either rejoining C at (a) the same node of departure or (b) a different node. If all such paths are restricted to type (a), one gets a contradiction to (10.1). Thus there is a path of type (b). If the path of type (b) is too long, one would get a cycle longer that C, a contradiction. Thus the length of the path must be restricted. But the restriction on this length also conflicts with (10.1). Thus C must in fact span D, hence D is hamiltonian.   □

Ghouila-Houri's theorem [G4] now follows readily.

Corollary 10.11a   If D is a strong digraph of order p such that $\deg v \geq p$ for all nodes v in D, then D is hamiltonian.   □

Another result which follows from Meyniel's Theorem is a result of Woodall [W7] which places a restriction on the sum of the outdegree of one node and the indegree of another.

Corollary 10.11b   Let D is a nontrivial digraph of order p. If every pair of distinct nodes u and v with u not adjacent to v satisfies

(10.2) $$od(u) + id(v) \geq p,$$

then D is hamiltonian.

Proof   In order to apply Meyniel's Theorem, we first show that D is strong. For arbitrary nodes u and v, we must show that v is reachable from u. If u is adjacent to v, we are done, so assume the contrary. Then (10.2) implies that there is a node w adjacent from u and adjacent to v. Hence v is reachable from u and D is strong.

Now for any two distinct nonadjacent nodes u and v of D, we have

$$\begin{aligned} \deg u + \deg v &= id(u) + od(u) + id(v) + od(v) \\ &= od(u) + id(v) + od(v) + id(u). \\ &\geq p + p > 2p - 1 \end{aligned}$$

Thus by Theorem 10.11, D is hamiltonian.   □

Corollary 10.11c   If D is a digraph of order p such that for all pairs of nonadjacent nodes u and v

$$(10.3) \hspace{3cm} \deg u + \deg v \geq 2p - 3,$$

then D has a spanning path.

Proof   First, (10.3) guarantees that D is at least weak. We can form a strong digraph D_1 as the symmetric join $D + K_1$ of a new node w to D, that is, add w and a symmetric pair of arcs between w and each node of D. For every pair of nodes u_1 and v_1 in D_1, we have

$$\deg u_1 + \deg v_1 \geq 2p - 3 + 4 = 2p + 1 = 2(p + 1) - 1.$$

As D_1 has order $p + 1$, Theorem 10.11 implies that D_1 has a hamiltonian cycle C. By deleting node w and its incident arcs in C, we obtain a spanning path of D. $\square$

Generalizations of Hamiltonian Digraphs

The first generalization of hamiltonian digraphs we consider are digraphs for which there is a spanning path from each node to each other node. A digraph D is *hamiltonian-connected* if there is a spanning $u - v$ path for all pairs of distinct nodes u and v in D. A hamiltonian-connected digraph is always hamiltonian, but the converse is not true as a directed cycle of order at least 4 shows. Overbeck-Larisch [O6] showed that by altering the bound in Woodall's Theorem (Corollary 10.11b) by only 1, hamiltonian-connectedness is guaranteed.

Theorem 10.12   Let D be a nontrivial digraph of order p. If every pair of distinct nodes u and v with u not adjacent to v satisfies

$$od(u) + id(v) \geq p + 1,$$

then D is hamiltonian-connected. $\square$

There is a natural analogy to the concept of a pancyclic graph. A digraph D of order p is *pancyclic* if D contains a directed cycle of each length k, $3 \leq k \leq p$. Thus, these digraphs are a special class of hamiltonian digraphs. Thomassen [T5] gave the following condition for a digraph to be pancyclic which is an analogue of Bondy's result for pancyclic graphs, see Theorem 4.16.

Theorem 10.13 Let D be a strong digraph of order $p \geq 3$ such that $\deg u + \deg v \geq 2p$ for all pairs u and v of nonadjacent nodes. Then D is either pancyclic or p is even and D is the digraph of $K_{p/2,p/2}$.   □

There are also results for strong digraphs which guarantee the existence of a path of a given length when the digraph might not have a spanning path. For example, Bermond, Germa, Heydemann, and Sotteau [BGHS1] showed that a strong digraph of order p with $id(v) \geq k$ and $od(v) \geq h$ for all v contains a path of length at least $\min\{h + k, p - 1\}$. Ayel [A4] showed that under the same assumptions a strong bipartite digraph contains a cycle of length at least $\max\{2k, 2h\}$.

EXERCISES 10.4

1. Let D be a digraph of order p such that for every pair of distinct nodes u and v with u not adjacent to v, $od(u) + id(v) \geq p - 1$. Then D contains a spanning path.

2. Let $k = \max_v \min\{id(v), od(v)\}$ in a digraph D without symmetric pairs of arcs. Then D has a path of length at least k.

3. If D is a strong digraph of order $p \geq 3$ such that $\deg u + \deg v \geq 2p+1$ for all pairs u and v of nonadjacent nodes, then D pancyclic.
   (Overbeck-Larisch [O6])

4. If D is a digraph of order $p \geq 3$ such that $od(v)$ and $id(v)$ are each at least $(p + 1)/2$ for all v in D, then D is pancyclic.

10.5   TOURNAMENTS

Perhaps the most studied digraphs are the tournaments. A *tournament* is a nontrivial oriented complete graph. All tournaments with two, three, and four nodes are shown in Figure 10.10. The first with three nodes is called a *transitive triple*, the second a *cyclic triple*.

In a round-robin tournament, a given collection of players or teams play a game in which the rules of the game do not allow for a draw, e.g., basketball but not chess. Every pair of players encounter each other and exactly one from each pair emerges victorious. The players are represented by nodes and for each pair of nodes an arc is drawn from the winner to the loser, resulting in a tournament.

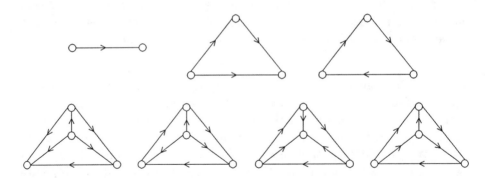

Figure 10.10   The seven smallest tournaments.

The first theorem on tournaments ever found is due to Rédei [R3]; for small tournaments it can be verified using Figure 10.10.

Theorem 10.14   Every tournament has a spanning path.

Proof   The proof is by induction on the number of nodes. Every tournament with 2, 3, or 4 nodes has a spanning path, by inspection. Assume the result is true for all tournaments with n nodes, and consider a tournament T with $n + 1$ nodes. Let v_0 be any node of T. Then $T - v_0$ is a tournament with n nodes, so it has a spanning path P, say $v_1 v_2 \ldots v_n$. Either arc $v_0 v_1$ or $v_1 v_0 \in T$. If $v_0 v_1 \in T$, then $v_0 v_1 v_2 \ldots v_n$ is a spanning path of T. If $v_1 v_0 \in T$, then let v_k be the first node (if any) of P for which the arc $v_0 v_k$ is in T. Then $v_{k-1} v_0 \in T$, so that $v_1 v_2 \ldots v_{k-1} v_0 v_k \ldots v_n$ is a spanning path of T. If no such node v_k exists, then $v_1 v_2 \ldots v_n v_0$ is a spanning path. In any case, we have shown that T has a spanning path, completing the proof. $\square$

By the way, Rédei [R3] also showed that every tournament has an odd number of spanning paths.

In Figure 10.3, we gave an example of a strong digraph with no spanning path. Theorem 10.14 shows that this cannot occur even for a weak tournament. Foulkes [F8] and Camion [C2] shows that a stronger statement holds for strong tournaments by characterizing them as follows.

Theorem 10.15   A tournament of order $p \geq 3$ is strong if and only if it has a spanning cycle. $\square$

Harary and Moser [HM6] showed that strong tournaments of order $p \geq 3$ are in fact pancyclic. Moon [M9] proved an even stronger result when he demonstrated that they are node-pancyclic, that is, for every node v, there is a cycle of each length k, $3 \leq k \leq p$, passing through v.

Score Sequences

Using terminology from round-robin tournaments, we say that the *score* of a node in a tournament is its outdegree and a node is said to *dominate* each node to which it is adjacent. The next theorem, due to Landau [L1], was actually discovered during an empirical study of tournaments (so-called "pecking orders") in which the nodes were hens and the arcs indicated pecking.

Theorem 10.16 In any tournament the distance from a node with maximum score to any other node is 1 or 2.

Proof Let v be a node with maximum score k in tournament T. Suppose u is a node at distance at least 3 from v. Then $uv \in T$ and u must dominate each of the k nodes that v dominates. Hence $od(u) \geq k + 1 > od(v)$, a contradiction.   □

In the same paper [L1], Landau characterized score sequences.

Theorem 10.17 A nondecreasing sequence of nonnegative integers $s_1, s_2, \ldots, s_p$ is the score sequence of a tournament T if and only if for each k, $1 \leq k \leq p$, we have

$$\sum_{i=1}^{k} s_i \geq \binom{k}{2},$$

with equality holding for $k = p$.   □

EXERCISES 10.5

1. The scores of a tournament satisfy $\sum s_i^2 = \sum (p - 1 - s_i)^2$.
2. Determine which of the following sequences are score sequences.
    a. $0, 1, 3, 3, 3$ b. $0, 1, 2, 3, 4, 4$ c. $1, 1, 1, 4, 4, 4$ d. $1, 2, 2, 3, 3, 5, 5$

3. State the dual of Theorem 10.16.

4. A tournament T is *transitive* if $uv \in T$ and $vw \in T$ implies that $uw \in T$. A tournament is transitive if and only if it is acyclic.

5. A tournament is *irreducible* if it is impossible to split the nodes into two sets V_1 and V_2 so that each node in V_1 dominates each node in V_2. A tournament is irreducible if and only if it is strong.

6. A sequence of numbers $s_1 \leq s_2 \leq \cdots \leq s_p$ is the score sequence of a strong tournament if and only if for all $k < p$,

$$\sum_{i=1}^{k} s_i > k(k-1)/2.$$

(Harary and Moser [HM6])

7. If no node in a tournament T has score $p - 1$, there are at least 3 nodes in T with eccentricity 2.

FURTHER RESULTS

Graham, Hoffman and Hosoya [GHH1] showed that the determinant of the distance matrix $\partial(W)$ of a weighted strong digraph W depends only on the distance matrices of the strong components and not on how the strong components are joined together. Let the blocks of W be $G_1, G_2, \ldots, G_n$, and let $\mathrm{cof}(M)$ denote the sum of the cofactors of M. Then

$$\det \partial(W) = \sum_{i=1}^{n} [\det \partial(G_i) \prod_{j \neq i} \mathrm{cof}\, \partial(Gj)].$$

1. If D is unilateral, then both $D - u$ and $D - v$ are unilateral.

(Fink [F3])

2. If A is the adjacency matrix of the line digraph of a complete symmetric digraph, then $A^2 + A = J$, the matrix with all unit entries. Illustrate this for $\mathbf{D}(K_3)$.   (Hoffman [H24])

3. Consider those digraphs in which for every node u, the sum $\sum d(u,v)$ of the distances from u is constant. Construct such a digraph which is not node-symmetric.   (Harary [H6])

CHAPTER 11

Graph Algorithms

Because of the importance of graph theory in computer science applications, the area of graph algorithms has been one of the fastest growing areas of graph theory. An *algorithm* is a step-by-step procedure for solving a problem. The general goal of algorithmic graph theory is to find an "efficient" algorithm for solving a given problem or to show that no such algorithm exists, where, loosely speaking, an efficient algorithm is one guaranteed to finish in an acceptably short period of time. When no such algorithm can be produced, one might instead try to find good heuristics (procedures guaranteed to have small run time and which will generally produce optimal or near-optimal solutions). Algorithmic graph theory has been such an active area of research that many books have already been written on the subject. We shall focus here on graph algorithms relating to the distance concepts we have discused throughout the text.

11.1 POLYNOMIAL ALGORITHMS AND $\mathcal{NP}$-COMPLETENESS

When searching for an efficient algorithm for a particular problem, one must first decide what *"efficient"* means. After deciding upon a definition, one would then like some way to measure efficiency. Such a measure could then be used for various purposes:

1. To compare different algorithms for the same problem.

2. To estimate the computer runtime for the algorithm.

3. To help decide if the algorithm is optimal.

To give a precise description of efficiency, we must first define several concepts. An *instance* of a problem is a particular example to which the algorithm is to be applied. For example, the Petersen graph is an instance of a graph. Each instance will be presented for solution in a particular format. For example, suppose we want an algorithm to determine whether a graph is bipartite. Each graph (instance) to be tested will be presented in the same way. One might suggest drawing the graph. Although this would work fine for small graphs, we want to use the algorithm on graphs of arbitrary size. Thus the graph is generally presented via some data structure. Some standard data structures used for graphs are arrays, linked lists, stacks, heaps, and packets. It should be noted that the choice of the data structure will affect both the efficiency and structure of the algorithm. Since we do not assume prior familiarity with data structures, we shall focus on the simplest ones, namely, arrays and lists. This is no disadvantage since most graph algorithms are first presented in these terms. Later, when refining the algorithm to increase efficiency more complicated structures might be used.

The simplest array used to describe a graph is its adjacency matrix (see §6.1). If the instance (graph) is not already labeled, one simply takes any fixed labeling and uses the adjacency matrix for that labeling. It should be noted that an adjacency matrix is not an efficient technique of describing the graph. Since for any graph G, $A(G)$ is symmetric, there is a built-in redundancy. Also, all diagonal elements of $A(G)$ are zero so are not needed. Thus the upper triangle of $A(G)$ would describe G just as well as all of $A(G)$. Hence one can completely described G with a sequence of $(p^2 - p)/2$ zeros and ones. The biggest such number among adjacency matrices for G is unique and characterizes G.

Another common technique of describing a graph is to use an adjacency list. Here one would need at most $(p^2 - p)/2$ entries, but usually considerably fewer. Begin by labeling the nodes $v_1, v_2, \ldots, v_p$. The *adjacency list* for node v_i consists of the nodes v_j for which $v_i v_j \in G$. The total number of entries on all the adjacency lists corresponds to twice the number of edges in G. This provides a considerable savings of computer storage when G is sparse.

An algorithmic procedure consists of reading the input string that describes the instance, processing the input string via the algorithm, and producing an output string. The whole procedure is preceded by coding

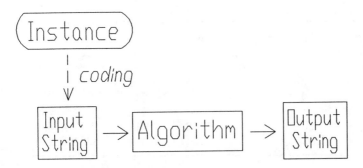

Figure 11.1   The algorithmic process.

the instance in the form of the input string. This process is depicted in Figure 11.1.

The *input length* is the number of elements in the input string. For the adjacency matrix this would be p^2, its number of 0's and 1's. For the collection of adjacency lists, the input length is $2q$. The efficiency of an algorithm is described as a function $f(n)$ of its input length n.

Orders of Magnitude

To understand efficiency, one needs some familiarity with the concept of orders of magnitude. When comparing two algorithms, the so-called *Landau notation* is often used. Suppose that f and g are functions defined on the natural numbers and that

$$\lim_{n \to \infty} f(n) = \lim_{n \to \infty} g(n) = \infty.$$

To say that $f(n) = O(g(n))$, read "f is big-oh of g," means that $f(n)$ grows no faster than $g(n)$ as $n \to \infty$. Hence, the growth of f is in some sense bounded by the growth of g. If $f(n) = O(g(n))$ and $g(n) = O(f(n))$, then each grows no faster than the other and f and g are said to have the *same order of magnitude*.

To say that $f(n) = o(g(n))$, read "f is little-oh of g," means that $g(n)$ grows much faster than $f(n)$ as $n \to \infty$. In this case, g is said to have a *higher order of magnitude* than f. Note that if $f(n) = o(g(n))$ then $f(n) = O(g(n))$. That is if f grows more slowly than g, it grows no faster than g.

The relation between orders of magnitude between f and g can be easily found by considering the limit $L = \lim_{n \to \infty} f(n)/g(n)$. If $0 < L < \infty$, then f and g are said to have the same order of magnitude. However, if $L = 0$ then $f(n) = o(g(n))$, and if $L = \infty$, then $g(n) = o(f(n))$.

Table 11.1 COMMON ORDERS OF MAGNITUDE

$\lg n$	n	$n \lg n$	n^2	n^3	2^n
0	1	0	1	1	2
1	2	2	4	8	4
2	4	8	16	64	16
3	8	24	64	512	256
4	16	64	512	4096	65536
5	32	160	1024	32768	4294967296

Let $\lg n$ denote $\log_2 n$. Some common orders of magnitude are n, $\lg n$, $n \lg n$, n^2, n^3, and 2^n. To help provide some insight into these orders of magnitude, some values for these functions are listed in Table 11.1.

Note that if one is far enough down in the table, each value in a given row is larger than the value to its left. Thus the six functions are arranged in the table with those having a higher order of magnitude further to the right. Remembering the following three basic facts helps considerably when comparing orders of magnitude.

1. A polynomial has the same order of magnitude as its leading term.
2. $\lg n = o(n)$, that is, $\lg n$ grows much slower than n does.
3. If $P(n)$ is a polynomial such that $\lim_{n \to \infty} P(n) = \infty$, and $f(n) = a \cdot b^{kn}$, where a, b, and k are all positive, then $P(n) = o(f(n))$. That is, an exponential function has a higher order of magnitude than any polynomial function.

The *complexity function* $C_A(n)$ of an algorithm A with input string of length n is the maximum number of basic operations that can be performed by A. Thus $C_A(n)$ is a measure of the worst case of the algorithm, that is, the slowest the algorithm can run. The *complexity* of an algorithm is the order of magnitude of its complexity function. Once the number of operations and the time required for each operation are known, the total time required by the algorithm can be computed. Hence, the complexity is sometimes called the *time complexity* of the algorithm.

An algorithm A is a *polynomial time algorithm*, or simply a *polynomial algorithm*, if there exists a polynomial $p(n)$ such that $C_A(n) = O(p(n))$. An algorithm that is not a polynomial algorithm is called an *exponential algorithm*. The *order of magnitude of an algorithm* is the order of magnitude of its complexity function. Thus, for example, an algorithm A is of order $n \lg n$ if $C_A(n) = n \lg n$.

Exponential algorithms should be avoided since the explosive growth of their complexity functions makes even moderately-sized problems intractable even with a super-fast computer. For example, using an algorithm A with $C_A(n) = 2^n$ on an computer that can do one comparison per microsecond ($1/10^6$ sec.) on an input string of length 40 would take $2^{40} = 1,099,511,627,776$ comparisons to complete the algorithm in the worst case. It would therefore take about $1,099,512$ seconds or about 13 days to complete the computation.

However, if we use a polynomial algorithm B with complexity function $C_B(n) = n \lg n$ that solves the same problem, the same computer would take about 0.0002 seconds to run. Thus it should be clear — avoid exponential algorithms when possible.

When choosing one algorithm over another, one must remember that orders of magnitude are meaningful only for large values of n. Thus, one can see from Table 11.1 that if the input length is small, for example $n = 8$, an exponential (2^n) algorithm could be more efficient than a polynomial (n^3) one.

$\mathcal{NP}$-Complete Problems

We now briefly consider a class of problems for which there is no known polynomial time algorithms. A complete discussion of these problems would fill a whole book. See, for example, Garey and Johnson [GJ1]. A formal description of $\mathcal{NP}$-complete problems will necessarily include a formal discussion of Turing machines; however, for the sake of brevity we shall be slightly less formal here.

A *decision problem* is a problem for which each instance is described with its input string and the answer to the problem will be either "yes" or "no." As the problem is being solved it is said to be in a particular "state." A Turing machine T is used to solve the decision problem by progressing from an initial state to the final state which will be either "yes" or "no." The other states we call *intermediate* states. A *deterministic* machine is one for which each intermediate state in the solution of an instance of the decision problem has only one *next* state.

There are two standard ways to describe a *non-deterministic* machine. One way says that some intermediate state in the solution has several possible next states. Another way of thinking of a non-deterministic machine is that its solution of a decision problem requires an initial guess (of which there are several possible candidates), after which each subsequent intermediate state will have a unique next state.

A problem is said to be in the class $\mathcal{P}$ if it can be solved in polynomial time on a deterministic machine for every instance of the problem. A nondeterministic machine is said to *solve* a decision problem DP if the following conditions hold for every instance of DP:

1. If DP is true, then there exists an initial guess that will cause the machine to end in state "yes".

2. If DP is false, then there is no initial guess that will cause the machine to end in state "yes".

Another way to state condition 2 is that when DP is false every initial guess would either lead to state "no" or the algorithm would not terminate.

A problem is in the class $\mathcal{NP}$ if every instance of the problem can be solved with a polynomial algorithm on a nondeterministic machine. (In case you have not already guessed, the $\mathcal{NP}$ stands for "nondeterministic polynomial".) It is clear that $\mathcal{P} \subset \mathcal{NP}$ since if $DP \in \mathcal{P}$, one could simply use *no guess* as the initial guess and then solve the problem on the nondeterministic machine in polynomial time. The outstanding problem in complexity theory is to determine whether $\mathcal{P} = \mathcal{NP}$.

Decision problem DP_1 has a *polynomial reduction* to DP_2 denoted $DP_1 \propto DP_2$ if

1. There is a function f mapping each input string I_1 of DP_1 to an input string $f(I_1) = I_2$ of DP_2 such that I_1 produces the answer "yes" for DP_1 if and only if $f(I_1)$ produces the answer "yes" for DP_2, and

2. There is a polynomial algorithm to compute $f(I_1)$.

Note that a polynomial solution for DP_2 would imply a polynomial solution for DP_1.

A problem DP_1 in $\mathcal{NP}$ is $\mathcal{NP}$-*complete* if for every DP_2 in $\mathcal{NP}$, $DP_2 \propto DP_1$. People often express this by saying that a problem in $\mathcal{NP}$ is $\mathcal{NP}$-complete if it is at least as hard as every other problem in $\mathcal{NP}$. There are other standard ways in which $\mathcal{NP}$-completeness are defined, which may or may not be equivalent to one another. There is also a concept of strong $\mathcal{NP}$-completeness. For additional details on these, see Garey and Johnson [GJ1] or Even [E4].

There are three basic steps used to show that a problem DP is $\mathcal{NP}$-complete:

1. Show that $DP \in \mathcal{NP}$.

2. Select some problem DP' that is known to be $\mathcal{NP}$-complete.

3. Construct polynomial reduction of DP' to DP.

Note that it is important to show that there is indeed a polynomial algorithm to compute the image $f(I)$ of an arbitrary instance $I \in DP'$ for the function f constructed in step 3.

Garey and Johnson describe six basic $\mathcal{NP}$-complete problems and use them throughout their book to show that numerous graphical problems are $\mathcal{NP}$-complete. Their clearly written text is an excellent tool for learning the techniques of showing that a given problem is $\mathcal{NP}$-complete.

EXERCISES 11.1

1. Suppose that the input string for a graph G of order p is its adjacency matrix A. You use the standard algorithm for matrix multiplication to find A^2.

   a. What is the input length?

   b. How many operations (multiplications and additions) are used?

   c. What is the complexity of the algorithm.

2. How much storage would be saved for the following graphs using adjacency lists rather than the adjacency matrix as the input string?
   a. Petersen graph   b. C_p c. $K_{m,n}$

3. Determine the complexity of the Algorithm 9.1 from Chapter 9, which determines whether a given sequence is a graphical degree sequence.

4. Suppose the input string for a problem consists of the sequence of eccentricities $e_i = e(v_i)$ of the nodes of a graph G. Describe an algorithm to determine the center. What is the complexity function of your algorithm (a basic operation here is a comparison)? What is the complexity of your algorithm?

5. Suppose you have two algorithms A_1 and A_2 for the same problem. Algorithm A_1 has complexity function $6n^3 \lg n$ and A_2 has complexity function 2^n. What is the largest value of n for which it makes sense to use A_2 rather than A_1?

6. If $DP_1 \propto DP_2$ and there is a polynomial algorithm solving DP_2, then there is a polynomial algorithm that solves DP_1.

7. The relation $\propto$ is transitive.

11.2 PATH ALGORITHMS AND SPANNING TREES

The most fundamental graph algorithms concerning distance are those dealing with shortest paths and longest paths in a graph. In fact, most algorithms involving distance use one of the two basic search techniques of graph theory — depth first and breadth first search, which we now discuss.

In most cases when describing algorithms, we shall assume that adjacency lists are used to describe any graph input to an algorithm. When using such an algorithm, if the graph is not already labeled, simply label the nodes and record the adjacency list for input. An example of a graph and its adjacency lists is given in Figure 11.2.

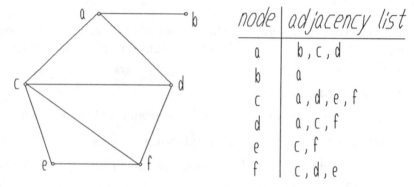

Figure 11.2 A graph and its adjacency lists.

Breadth-First Search

Breadth-first search (BFS) is a fundamental search technique that traces a rooted spanning tree in a connected graph so that the distance from the root to each node in the tree corresponds to its distance in the original graph. The basic idea is to begin at the root and find its neighbors, and then their neighbors, and so on, until one has spanned throughout the graph and reached all nodes. Since in most applications where BFS is used, one wants to know the distance from the root to each other node, we present a form of the algorithm which records the distances. In fact the version of the algorithm we present is fairly general in that it discovers whether the graph is connected rather than assuming that it is. The input to the algorithm is the list of nodes, their adjacency lists, and the label of the root. Assume that the distance $d(root, v_k)$ from the root to node v_k is stored in array d. Let $N(v)$ denote the adjacency list of node v.

Algorithm 11.1   Breadth-First Search
begin
    $d(root) = 0,\quad i = 1 \quad D = \{root\}$
    $C = N(root)$ (records current neighborhood being processed)
    while $D \neq \emptyset$ **do**
    begin
        $i = i + 1$
        $C = D$
        $D = \emptyset$ (D accumulates next neighborhoods)
        for each $w \in N(C)$ **do**
        begin
            place w in D
            remove w from all adjacency lists
            $d(w) = i$
        end
    end
    for each v not yet labeled, $d(v) = \infty$.
    If such v exist, the graph is disconnected, and all
    such v lie in a different component from the root
end   □

When first looking at this algorithm, it may seem that a lot of searching occurs when we try to remove a node w from all the adjacency lists. Actually, this step is quite simple, since $N(w)$ tells precisely the lists from which w must be removed. The complexity of BFS on a (p, q)-graph is $O(q)$. Since $q = O(p^2)$, BFS is quite an efficient method for finding the distance from one node to all the other nodes in the graph, and is extremely effective for sparse graphs.

By performing a BFS from each node, one can determine the distance from each node to each other node. The resulting algorithm has complexity $O(pq) = O(p^3)$. Note that it is also quite easy to modify Algorithm 11.1 with a labeling scheme to recover a shortest path (rather than just its length) from the root to each node without increasing the complexity. In this way we can generate a spanning tree where the distance from each node to the root in the tree is the same as in the original graph.

Depth-First Search

For a *depth-first search (DFS)* begin at the root and trace out a path from the root until you can go no farther without revisiting a node. Then backtrack along the path until reaching the first node with an alternate route available and procede forward again. Repeat this until you can go no farther. Like BFS, a depth-first search traces a spanning tree in a connected graph, but in a different manner. One might think of DFS as the way an intelligent but determined mouse might find its way through a maze.

A *procedure* is a self-contained algorithm used by another algorithm that "calls it." A *recursive procedure* is a procedure that calls itself during processing. Because of the repetitive nature of DFS, it is often written as a recursive algorithm (this is an algorithm using a recursive procedure).

Algorithm 11.2   Depth-First Search
Procedure DFS(v)
begin
    $l(v) = i$ (tracks the order in which nodes are visited)
    $i = i + 1$
    while $N(v) \neq \emptyset$ **do**
      for $u \in N(v)$ **do**
      begin
        $T = T \cup \{u, v\}$
        remove u from all adjacency lists
        DFS(u)
      end
end DFS

begin   (driver algorithm)
    input adjacency lists and *root*
    $T = \emptyset$ (stores edges of the tree as they are selected)
    for $v \in G$ **do**
      $l(v) = 0$
    $l(root) = 1$
    $i = 2$
    while there exist some u for which $l(u) = 0$ **do**
    for highest labeled node v with $N(v) \neq \emptyset$ **do**
    begin
      remove v from all adjacency lists
      DFS(v)
    end
    print T.
end   □

The backtracking along a path in the tree occurs each time control is returned to the driver program and one looks for the highest labeled node that has unvisited neighbors. Note that as with BFS, removing a node w from all adjacency lists in this algorithm is easy because $N(w)$ tells which lists are involved. Algorithm 11.2 can easily be modified to determine whether G is connected simply by testing for nodes v that still have $l(v) = 0$ when the algorithm terminates.

The DFS and BFS algorithms are used in numerous problems such as checking connectivity, determining the blocks or finding bridges in a graph, and determining distances between nodes. For example, Tarjan [T2] showed that one can determine whether a graph is biconnected in $O(p^2)$ time by using DFS. Ebert [E1] used that result to develop an algorithm linear in q to find a pair of node-disjoint paths joining any pair of nodes in a biconnected graph. BFS and DFS are also the basis of various algorithms on digraphs and networks. We shall discuss network algorithms in Chapter 12.

Some Comments on Spanning Trees

The following is a recent example on the use of BFS for graphs. Theorem 7.20 gives a characterization of graphs G having a diameter-preserving spanning tree (DPST), that is, a spanning tree T whose diameter is the same as the diameter of G. The proof in [BL1] is constructive and generates the DPST when the conditions of the theorem are satisfied. When $d = 2r$, the DPST is obtained by performing a BFS in G using a central node as root. When $d = 2r - 1$, a graph G^* is first formed from G by collapsing an edge joining a pair of central nodes. A BFS is then used to produce a DPST T^* of G^*, and after some simple manipulations of T^* a DPST of G is obtained.

Although BFS and DFS each produce a spanning tree in a connected graph G, the resulting trees are often quite different. Either search technique might produce several trees, but those produced from BFS generally will be more compact and have smaller diameter than those from DFS. For example, consider the Petersen graph P. Since P is node-symmetric, it is immaterial which node is used as the root. With BFS a tree isomorphic to that in Figure 11.3a is obtained, whereas DFS produces a tree isomorphic to one of the three trees in Figure 11.3b. Hence, BFS on P produces a tree of diameter 4, while DFS produces a tree with diameter 8 or 9.

The *height* of a rooted tree is the maximum distance of a node from the root. Like BFS, the complexity of DFS is $O(q)$ and $O(p^2)$. However, Fellows, Friesen and Langston [FFL1] showed that the problems of finding a spanning tree of maximum height or a spanning tree of minimum height using DFS are each $\mathcal{NP}$-hard. The basic difficulty is that for a given

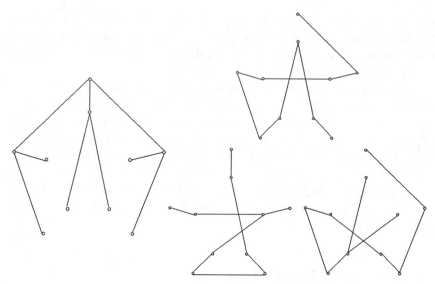

Figure 11.3   Spanning trees of the Petersen graph from BFS and DFS.

choice of root, DFS will not necessarily generate the desired "tallest" or "shortest" tree on the first try. Indeed, a DFS may have to be performed many times to obtained such a tree.

Define the *distance $d(T_1, T_2)$ between two spanning trees T_1 and T_2 of a graph G* to be the number of edges in T_1 that are not in T_2. It is easy to show that this is a metric on the set of all spanning trees of G. Two spanning trees T_1 and T_2 of G are *maximally distant* if $d(T_1, T_2)$ is at least as large as the distance between any pair of spanning trees of G. This concept was studied by Ishizaki, Ohtsuki, and Watanabe [IOW1]. Suppose that N is a network with underlying graph G. The maximum distance between a pair of spanning trees of G corresponds to the minimum number of independent variables required in the "mixed-variable analysis" of N. See Swamy and Thulasiraman [ST2] for a further discussion of this analysis technique.

EXERCISES 11.2

1. Show all steps in BFS by applying Algorithm 11.1 to the graph in Figure 11.4.

2. Do Exercise 1 using DFS and Algorithm 11.2.

3. When BFS terminates, the value of $d(v)$ is the distance from the root to v for each node v.

4. If we want DFS to generate a spanning forest when G is disconnected, indicate precisely which statements must be modified in Algorithm 11.2, and show the required changes.

5. Verify that the distance $d(T_1, T_2)$ between two spanning trees T_1 and T_2 of a graph G is a metric on the set of all spanning trees of G.

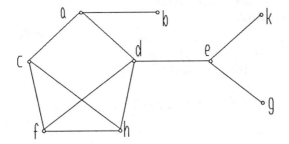

Figure 11.4   A graph to search.

11.3   CENTERS

Algorithms have been developed for many of the centrality concepts discussed in Chapter 2. In this section, we present several such algorithms.

The Center

To determine the center itself, one must know the eccentricities of all the nodes and select those nodes v for which $e(v)$ is minimum. There are a couple of standard ways to do this. We shall present here techniques one might use for a graph G. When dealing with a network where there are either weights on the edges or both weights and directions, other techniques are usually more appropriate and efficient. We shall discuss algorithms used to find the center of a network in §12.5.

Perhaps the most elementary way to determine the center uses the adjacency matrix $A(G)$ as input and produces the distance matrix simply by calculating consecutive powers of $A(G)$ as in [HNC1]. By Corollary 6.1c, $d(v_i, v_j)$ for $i \neq j$ is the least integer n for which $[A^n]_{ij} \neq 0$ when G is connected. When G is disconnected, $d(i, j)$ has the same value n when v_i and v_j are in the same component of G. However, when v_i and v_j are in different components, $[A^n]_{ij} = 0$ for all n. In this case, $d(v_i, v_j) = \infty$. Theorem 6.1 asserts that $[A^n]_{ij}$ is the number of walks of length n from v_i to v_j, and since $d(G) \leq p - 1$ when G is connected, this technique of determining the eccentricities could require us to calculate A^{p-1} in the worst case.

The algorithm producing eccentricities by using the distance matrix $D = D(G)$ obtained from the adjacency matrix $A = A(G)$ follows.

Algorithm 11.3   Eccentricities I
Procedure DfromA
begin
    $k = 1; \quad B(1) = A; \quad D = A$
    for each $i \neq j$ with $d_{ij} = 0$ **do**
      $d_{ij} = \infty$
    while some $d_{ij} = \infty$ and $k < p - 1$ **do**
    begin
      $k = k + 1$
      $B(k) = A \cdot B(k - 1)$
      for each ij where $d_{ij} = \infty$ **if** $[B(k)]_{ij} \neq 0$ **do**
        $d_{ij} = k$
    end
end DfromA
begin   (driver algorithm)
    input p and A
    DfromA
    for each i **do**
      $e_i = \max_j d_{ij}$
    print all e_i
end $\square$

Using Algorithm 11.3, it is trivial to determine the center. Simply locate the nodes with minimum eccentricity and note that

$$C(G) = \{v_i : e_i \leq e_j, \ i \leq j \leq p\}.$$

Although using Algorithm 11.3 is not the most efficient way to find the center, it is still quite popular since it is so easy to program. Indeed, most students in a first semester programming course would be able to write the code for Algorithm 11.3 once they covered arrays and procedures. In fact, with slight modification, one could easily program this using arrays but no procedures.

Algorithm 11.3 could involve many multiplications, however we focus on the more expensive comparison operations to measure efficiency. The main input is A which has input length $n = p^2$. Initially there are $O(n)$

comparisons to find the zero locations that should be initialized to ∞. It may seem that there are a huge number of comparisons to find which positions are still ∞ after each calculation of $B(k)$. However, this is not necessarily the case. The positions of matrices D and B can be labeled $1, 2, \ldots, p^2$ so that d_{ij} and b_{ij} are labeled $(i-1)p + j$. The labels of locations of ∞ are stored in a list L. Then after each calculation of a new $B(k)$, simply run through the list L and check the corresponding entries of $B(k)$. For each entry b_t that is no longer zero, adjust d_t and delete t from the list L.

Another standard way of determining all eccentricities, and thereby finding the center begins with adjacency lists and uses BFS repeatedly with each node successively playing the role of root. Assume we have BFS modified to a procedure whose input is the root and whose output is the eccentricity of the root. Thus Algorithm 11.1 is modified to a procedure BFS($root, d(root)$) by changing its final end statement to "end BFS" and preceding it by the extra step

> **if** $d(v) = \infty$ for some v **then**
> $e_{root} = \infty$
> **else**
> $e_{root} = \max d(v)$

Using BFS discovers with the first node whether G is connected and thus whether all nodes will have eccentricity ∞. So in that case, BFS is only invoked once and all nodes are then known. We now present the algorithm to determine eccentricities using the recursive BFS procedure.

Algorithm 11.4   Eccentricities II
begin
   input p and adjacency lists for each v_j, $1 \leq j \leq p$
   $j = 1$
   BFS(v_j, e_j)
   if $e_j = \infty$ **then**
      For $k = 2$ to p do
         $e_k = \infty$
   else while $j < p$ **do**
   begin
      $j = j + 1$
      BFS(v_j, e_j).
   end
end   □

Now finding the center or periphery of G is simply a matter of checking the eccentricities found by the algorithm.

The Median

To find the median of a graph, we modify Algorithms 11.1 and 11.4 to determine the status values rather than eccentricities. It consists of a procedure BFSS to find the status of a given node. This procedure is called by the driver program for each node in G.

Algorithm 11.5   Status
Procedure BFSS($root, s(root)$)
begin
 $s(root) = 0$; $i = 0$
 $D = \emptyset$ (accumulates next neighborhoods)
 $C = root$ (records current neighborhood being processed)
 $V = V(G) - \{root\}$ (tracks nodes not yet visited)
 while $C \neq \emptyset$ **do**
 begin
  $i = i + 1$
  for each $w \in N(C)$ **do**
  begin
   $D = D \cup \{w\}$
   remove w from V and all adjacency lists
   $s(root) = s(root) + i$.
  end
  $C = D$
 end
 if $V \neq \emptyset$ **then** $s(root) = \infty$
end BFSS
begin
 input p and adjacency lists for each v_j, $1 \leq j \leq p$
 $j = 1$
 BFSS($v_j, s(v_j)$)
 if $s(v_1) = \infty$ **then**
 begin
  for $k = 2$ **to** p **do**
   $e_k = infty$
 end
 else while $j < p$ **do**
 begin

$$j = j + 1$$
$$\mathbf{BFSS}(v_j, s(v_j))$$
    end
end   □

After completion of Algorithm 11.5, the status of each node is known, hence the median is immediately available. This algorithm has efficiency $O(p^3)$, the same as the algorithm for determining all eccentricities. Note that if it is known that the graph is a tree T, both the center and the median can be found in linear time by taking advantage of the following:

1. Paths between any given pair of nodes in T are unique (Theorem 1.2).

2. The center $C(T)$ consists of a single node or a pair of adjacent nodes (Theorem 2.1).

3. The median of a tree consists of a single node or a pair of adjacent nodes (Theorems 2.3 and 2.11).

4. For any path $v_1, v_2, \ldots, v_k$ joining two endnodes of T, the values of both e_i and $s(v_i)$ decrease as we move in from the ends of the path.

Generalized Centers

Centrality concepts other than the center and median have received far less attention in terms of algorithmic results. In most cases, algorithms have been developed for the class of trees but not for more general classes of graphs. We shall only present one such algorithm here, but first we develop a procedure that is useful in several distance algorithms. This procedure determines whether a given graph is a path.

Procedure PATH$(p, G, path)$
begin
    $j = 1$; $path = true$
    if there exists v such that $\deg v = 1$ **then**
    begin
        $root = v$
        while $|N(root)| = 1$ **do**
        begin
            $temp = u$ in $N(root)$
            $N(u) = N(u) - \{root\}$
            $root = temp$
            $j = j + 1$
        end

    else if $p = 1$ **then** $root = v$
    if $j = p$ **and** $|N(root)| = 0$ **then**
      $path = true$
    else
      $path = false$
end PATH   □

Restricting attention to trees sometimes results in algorithms that are quite simple. For example, as mentioned in §2.4, Cockayne, Hedetniemi and Hedetniemi [CHH1] showed how to find the path center of a tree.

Algorithm 11.6   Path Center of a Tree
begin
    input p and adjacency lists for each v_j, $1 \le j \le p$
    PATH($p, G, path$)
    while $path = false$ **do**
    begin
      for each v with $|N(v)| = 1$ **do**
      begin
        remove v from all adjacency lists and from V
        $p = p - 1$
      end
      PATH($p, G, path$)
    end
end   □

Algorithms to determine various other types of generalized centers of trees were developed by Slater [S19,19a], and Morgan and Slater [MS3]. Also, some additional algorithms are presented in Handler and Mirchandani [HM1] and Minieka [M8]. In most cases, algorithms for generalized centers of more general classes of graphs are not yet available.

EXERCISES 11.3

1.   Apply Algorithm 11.4 to the graph in Figure 11.4.
2.   Do Exercise 1 using Algorithm 11.5.
3.   Use Algorithm 11.6 to find the path center of the tree in Figure 11.5.
4.   Determine the efficiency of Algorithm 11.6.

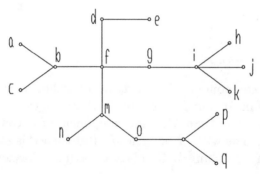

Figure 11.5   A graph to center attention on.

11.4   MAXIMUM MATCHINGS

In §3.3 we saw that in a bipartite graph G the size of a maximum matching equals the minimum number of nodes required to cover all the edges of G. It is easy to see that this statement cannot be extended to nonbipartite graphs simply by considering any odd cycle. In this section we present an algorithm for finding a maximum matching in an arbitrary graph. When G is bipartite, another algorithm using network flows is more common. We shall discuss that algorithm in Chapter 12.

In order to describe an algorithm for maximum matchings, we first need some additional terms. Let $M \subset E(G)$ be a matching of G. An edge uv in M is said to be *matched* as are nodes u and v. A node that is not incident with an edge of M is *unmatched*. A path P is an *alternating M-path* if exactly one of each pair of consecutive edges of P is in M. An *M-augmenting path* is an alternating path that begins and ends at distinct unmatched nodes. Berge [B4] characterized maximum matchings in graphs.

Theorem 11.1   A matching M in G is a maximum matching if and only if there is no M-augmenting path in G.

Proof   Clearly every maximum matching has no M-augmenting path P, since otherwise deleting the matched edges of P from M and adding the other edges of P to M would produce a larger matching of G.

For the converse, suppose that M is a matching that is not maximum, and let M' be a maximum matching of G. We will show that there is an M-augmenting path. Consider the minimum order subgraph H of G whose edge set is $(M - M') \cup (M' - M)$, that is, the edges in exactly one of the matchings. A node in H is incident with either

1. one edge of M,

2. one edge of M', or

3. one edge of both M and M'.

Thus each component of H consists of either an even cycle whose edges alternate in M and M' or a path whose edges alternate in M and M'. Since M' is larger than M, some component of H consists of a path P that begins and ends with an edge in M'. But then P is an M-augmenting path. Hence, any matching M that is not maximum has an M-augmenting path.   □

Virtually all algorithms to find maximum matchings are based on Theorem 11.1. They generally begin with a matching (or the null matching) and continually look for an augmenting path to get a larger new matching until no augmenting path can be found. Since the process is so repetitive and the number of paths grows exponentially with the size of the graph, the procedure for finding an augmenting path becomes the key to making a particular algorithm efficient. Edmonds [E2] developed a technique that made this problem feasible.

The process begins by finding an unmatched node and adding an edge incident with it to the matching. Continue to find such a node v and add such an incident edge, if possible. Otherwise, a tree is generated to find an alternating path beginning at v and ending at another unmatched node. In Edmonds' version, when certain odd cycles (called "blossoms") are discovered, they are contracted to a node and the process continues. However, when an alternating path joining v with another unmatched node is discovered, some extra work is necessary to track the actual path if it passes through a "blossom node."

The efficiency of Edmonds' algorithm is $O(p^4)$. By using certain labeling techniques, various authors have improved the efficiency. For example, Gabow [G1] uses labels in a variation of Edmonds' method and thereby avoids the contraction process and achieves efficiency $O(p^3)$. Swamy and Thulasiraman [ST2] contains a detailed discussion of that approach. We shall discuss instead a technique of Conradt and Pape [CP2]. Rather than beginning with the empty matching, quickly generate a reasonable-sized matching by making one pass through the adjacency lists and pairing off unmatched nodes. In the next procedure, *partner* is a p-vector, and, as usual, G is input in terms of the adjacency lists.

Procedure MATCH$(p, G, unmatched, partner)$
begin
    $unmatched = p$
    for $j = 1$ to p **do**
      $partner(j) = 0$
    for $j = 1$ to p **do**
      if $N(v_j) \neq \emptyset$ **then**
      begin
        for the first v_k listed in $N(v_j)$ **do**
        begin
          $partner(j) = k$
          $partner(k) = j$
        end
        delete v_j and v_k from all adjacency lists
        $unmatched = unmatched - 2$
        $N(v_k) = \emptyset$
      end
end MATCH   □

After this procedure is used, we have a good-sized matching with which to start and the vector *partner* tells which nodes are paired in the matching. Any node v_j for which $partner(j)$ is still 0 remains unmatched. Also, the variable *unmatched* keeps track of how many unmatched nodes there are. Thus, if *unmatched* is 0 or 1 after the procedure is called, we have a maximum matching. Otherwise, we try to generate an alternating path between a pair of unmatched nodes. This is accomplished when possible by generating a rooted tree beginning at an unmatched node (the root). For any pair of adjacent edges of the tree, exactly one is matched. Hence, all paths in the tree are alternating paths. During the process, nodes are labeled "*odd*" or "*even*" according to whether their distance from the root in the tree is odd or even. Thus we get an alternating tree as shown in Figure 11.6 where unmatched edges are dotted.

A *queue* is a data structure for which the earlier an item is placed on the queue the earlier it will leave. We use a queue called *unused* to store even endnodes. A *boolean array* is a vector each of whose entries is either 0 or 1. We use a boolean array b to keep track of the root and odd nodes. The algorithm includes a procedure to augment the matching when an M-augmenting path is found. This procedure alters the array partner that is passed to it and describes M. The values i, j passed to the procedure are the subscripts of the two final nodes of the M-augmenting path.

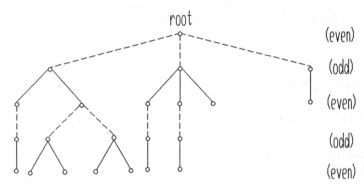

Figure 11.6   An alternating tree.

Algorithm 11.7   Maximum Matching
Procedure AUGMENT$(p, G, partner, back2, i, j)$
begin
    while $partner(i) \neq 0$ **do**
      $partner(j) = i$
      $partner(i) = j$
      $i = back2(i)$
      $j = back2(j)$
    until $partner(i) = 0$ (the root has been reached)
    $partner(j) = i$
    $partner(i) = j$
end AUGMENT
begin
    input p and G
    MATCH$(p, G, unmatched, partner)$
    for $i = 1$ to p **do**
    begin
      if $unmatched \geq 2$ and $partner(i) = 0$ **then**
      begin
        for $j = 1$ to p **do**
          $b(j) = 0$
        $b(i) = 1$ (this is the root)
        Place v_i on queue $unused$
        $path = false$ (no alternating path yet found)
      end
      while $(unused \neq \emptyset$ and $path = false)$ **do**
      begin
        remove next node v_k from $unused$
        while $path = false$ **do**

```
      begin
        for each v_j ∈ N(v_k) do
          if b(v_j) = 0 and partner(j) = 0 then
          begin
            path = true
            AUGMENT(p, G, partner, back2, k, j)
            unmatched = unmatched − 2
          end
          else if partner(j) ≠ i then
          begin
            next = partner(j)
            if k ≠ i or next is not ancestor of v_k then
            begin
              b(j) = 1
              back2(next) = k
              place next on queue unused
            end
          end
        end
      end
    end
  end
end                                                           □
```

Algorithm 11.7 has efficiency $O(p^3)$, so the technique of Conradt and Pape [CP2] improves on the original algorithm of Edmonds by a a factor of p. The fastest known algorithm for maximum matchings in graphs is due to Micali and Vazirani [MV1]. A computer implementation of Conradt and Pape's algorithm is given in Sysło, Deo, and Kowalik [SDK1]. They also describe numerous other algorithms including algorithms specialized for the case when G is bipartite. Matchings for bipartite graphs are often given separate treatment because of their application in assignment problems (such as assigning a group of employees to jobs for which they are qualified). The bipartite case is also given extensive coverage in Gould [G10].

EXERCISES 11.4

For Exercises 1-3 recall that a *perfect matching* of a graph is a matching that covers all its nodes.

1. A tree has at most one perfect matching.

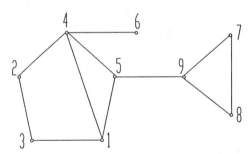

Figure 11.7   A graph to trace through.

2.   Characterize the complete multipartite graphs $K(n_1, n_2, \ldots, n_k)$ that have a perfect matching.

3.   Every cube Q_n $(n \geq 2)$ has a perfect matching.

4.   Find the initial matching for the graph in Figure 11.7 using the procedure MATCH.

5.   Apply Algorithm 11.7 to the graph in Figure 11.7.

6.   Since one need not worry about odd cycles, algorithms for maximum matchings in bipartite graphs are generally much easier to describe. State an algorithm for finding a maximum matching in a connected bipartite graph.

11.5   TWO $\mathcal{NP}$-COMPLETE PROBLEMS

Since there are no known polynomial algorithms to solve $\mathcal{NP}$-complete problems, they must be handled in a special manner. Some of the approaches used are as follows:

1.   Branch-and-bound techniques, which limit the amount of searching one must do. The algorithm is still exponential for the worst case but gives a drastic improvement for the average case.

2.   Approximation techniques, which produce a result close to the optimum (and occasionally produce an optimum solution) in polynomial time. In some situations, the sacrifice of obtaining a good approximation may far outweigh the extra time necessary to find a true optimum solution.

3.   Limit the problem to a domain for which a polynomial algorithm can be obtained.

In this section, we discuss two $\mathcal{NP}$-complete problems: finding a maximum clique in a graph and finding a longest path or cycle in a graph.

Maximum Cliques

One of the six basic $\mathcal{NP}$-complete problems in Garey and Johnson [GJ1]
is to determine for a given integer k whether a graph has a clique of order
at least k. A most efficient yet necessarily nonpolynomial algorithm for
generating *all* cliques of a given graph is presented in Reingold, Nievergelt,
and Deo [RND1]. Of course, if you just want to know the order of a largest
clique, you need not generate all cliques. Nevertheless, the problem is
still $\mathcal{NP}$-complete. Thus you may decide to settle for an approximation
algorithm. A standard assumption you might make in the approximation
algorithm is that a maximum clique contains a node of maximum degree
(which, of course, might not be the case). Thus you build a (hopefully
largest) clique around a node of maximum degree. The following is a
polynomial algorithm to accomplish this.

Algorithm 11.8   Big Clique

Procedure DEGREE(p, M, deg[array of degrees])
begin
    for $j = 1$ to p **do**
    begin
      $deg(j) = 0$
      while $M(v_j) \neq \emptyset$ **do**
      begin
        $deg(j) = deg(j) + 1$
        remove next node from adjacency list $M(v_j)$
      end
end DEGREE

Procedure MAXDEG(V, deg, x)
begin
    $t = 0$
    while $V \neq \emptyset$ **do**
    begin
      remove some v_k from V
      $V = V - v_k$
      if $deg(k) \geq t$ **then**
      begin
        $t = deg(k)$
        $x = k$
      end
    end
end MAXDEG

begin
    input p, V, and adjacency lists $N(j)$ for each v_j
    DEGREE(p, N, deg)
    MAXDEG(V, deg, x)
    $clique = \{v_x\}$
    let V be the set of nodes in $N(v_x)$
    while $V \neq \emptyset$ **do**
    begin
      $p = |V|$
      for each $v \in V$ **do**
        remove nodes not in V from $N(v)$
      DEGREE(p, N, deg)
      MAXDEG(V, deg, x)
      $clique = clique \cup \{v_x\}$
      let V be the set of nodes in $N(v_x)$
    end
end   □

During each pass through the while-loop of the driver program, a node with the most adjacencies to the remaining nodes is selected and placed in *clique*.

Hamiltonian Cycles

The problem of determining whether G has a hamiltonian cycle is also $\mathcal{NP}$-complete. Hence, there is little hope of finding a polynomial algorithm for it. What strategy should we then use. Since there are so many sufficient conditions to show that a given graph is hamiltonian, we shall alter the problem to that of actually finding a hamiltonian cycle in a graph that is known to be hamiltonian by some criterion such as

(11.1)   The closure $cl(G)$ is complete.

    or

(11.2) $\deg u + \deg v \geq p$ for all pairs of nonadjacent nodes.

In such cases we can find a hamiltonian cycle in polynomial time. Suppose that (11.1) holds. Our next algorithm is inspired by Bondy and Chvátal [BC1]. In the algorithm, A is the adjacency matrix and *deg* is a vector storing the degrees of the nodes.

Algorithm 11.9   Hamilton I
Procedure CLOSURE(p, A)
begin
    $zeros = \emptyset$
    for $i = 1$ **to** p **do**
    begin
      $deg(i) = 0$
      for $j = 1$ **to** p **do**
      begin
        if $A(i,j) = 0$ **then**
          if $i < j$ **then** $zeros = zeros \cup \{(i,j)\}$
        else
          $deg(i) = deg(i) + A(i,j)$
      end
    end
    $m = 2$
    while some $(i,j) \in zeros$ has $deg(i) + deg(j) \geq p$ **do**
    begin
      $A(i,j) = m$
      $A(j,i) = m$
      $deg(i) = deg(i) + 1$
      $deg(j) = deg(j) + 1$
      $m = m + 1$
    end
end CLOSURE
Procedure LARGELABEL($p, C, A, large, index$)
begin
    $large = A(C(p), C(1))$
    $index = p$
    for $j = 1$ **to** $p - 1$ **do**
      if $A(C(j), C(j+1)) > large$ **do**
      begin
        $large = A(C(j), C(j+1))$
        $index = j$
      end
end LARGELABEL
begin
    input p and A
    CLOSURE(p, A)
    for $i = 1$ **to** p **do**
      $C(i) = i$

$C(p+1) = 1$
LARGELABEL($p, C, A, large, index$)
while $large > 1$ **do**
begin
   cyclically shift C upward so that $C(index)$ is in $C(p)$
   find k so that $\max\{A(C(1), C(k+1)); A(C(2), C(k))\} < large$
   $D = C$
   for $j = 2$ to $p - k + 1$ **do**
     $C(j) = D(k + j - 1)$
   for $j = p - k + 2$ to p **do**
     $C(j) = D(p - j + 2)$
   LARGELABEL($p, C, A, large, index$)
  end
end   □

The efficiency of the procedure CLOSURE is $O(p^4)$, whereas the remainder of the driver program (done independently) is $O(p^3)$. However, since the whole process depends on the edge labels m generated in CLOSURE, we should say that the the efficiency of Algorithm 11.8 is $O(p^4)$. Remember that (11.1) must hold before applying Algorithm 11.1. Note that (11.2) implies (11.1), so with (11.2) we could use use the same algorithm. However, Albertson [A2] discovered the following $O(p^2)$ algorithm which applies when (11.2) holds.

Algorithm 11.10   Hamilton II
Procedure MAXIMALPATH($p, M, path, n, pass$)
begin
   $i = path(n)$
   delete all nonzero k in $path$ from all $M(j)$ except $M(1)$
   while $M(i) \neq \emptyset$ **do**
   begin
     remove some t from $M(i)$
     $n = n + 1$
     $path(n) = t$
     delete t for all $M(j)$
     $i = t$
   end
   $n = n - 1$
   reverse the first n entries of $path$
   if $pass = 1$ **then**

MAXIMALPATH$(p, M, path, n, 2)$
end MAXIMALPATH
begin
   input p and adjacency lists $N(v_j)$
   let $M(j)$ be the set of subscripts of nodes in $N(v_j)$
   for $j = 1$ **to** p **do**
     $path(j) = 0$
   $path(1) = 1$
   $n = 1$
   MAXIMALPATH$(p, M, path, n, 1)$
   if $n = p$ **and** $path(n) \in N(v_1)$ **then**
     $path, path(1)$ **describes a hamiltonian cycle**
   else while $n < p$ **do**
   begin
     if $path(n) \in N(v_1)$ **then**
       $path, path(1)$ **describes a cycle** C
     else
       find k such that $path(1) \in N(v_{k+1})$ and $path(n) \in N(v_k)$
     $C = v_{path(1)}, v_{k+1}, v_{k+2}, \ldots, v_{path(n)}, v_k, v_{k-1}, \ldots, v_{path(1)}$
     find $v^* \in G - C$ **adjacent to a node in** C
     determine a path P^* **containing** v^* **and every node of** C
     list consecutive subscripts of P^* **in vector** $path$
     $n = n + 1$
     MAXIMALPATH$(p, M, path, n, 1)$
   end
end   □

Note that MAXIMALPATH is a recursive procedure. It calls itself in order to generate a maximal path by expanding from the other end of the given path, if possible.

EXERCISES 11.5

1. Use Algorithm 11.8 to find a maximum clique in graph G of Figure 11.8.

2. Construct a connected graph on 7 nodes for which Algorithm 11.8 fails to find the maximum clique.

3. Use the procedure CLOSURE or Algorithm 11.9 to show that the closure of graph H in Figure 11.8 is K_6. What label m does edge $(1,4)$ in $cl(H)$ receive?

4.   Apply Algorithm 11.9 to graphs G and H of Figure 11.8.

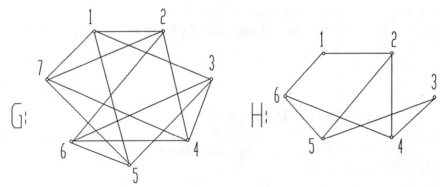

Figure 11.8   Graphs to cycle through.

CHAPTER 12

Networks

The analysis of many problems in transportation, telecommunications, and operations research requires a structure that contains more information than a graph or digraph. The additional information required could be the distance between pairs of neighboring cities, transmission times between switching centers, cost of transporting an item along a given route, the flow capacity of a pipe, or the impedance of a wire. In this chapter, we discuss techniques used to handle many such problems. In the process we extend many of the distance concepts we have discussed throughout the text and develop several additional algorithms.

12.1 THE MAX-FLOW MIN-CUT THEOREM

A *two terminal network* $\mathcal{N}$ consists of a digraph D having two distinguished nodes s and t together with a nonnegative real-valued function c defined on the arcs of D. Node s is called the *source* and t the *sink* of the network; the other nodes of $\mathcal{N}$ are the *intermediate nodes*. In general, a network need not have terminals. For each arc $e \in \mathcal{N}$, the value $c(e)$ is the *capacity* of e.

A *(legal) flow* f in $\mathcal{N}$ is an assignment of nonnegative real numbers $f(e)$ to the arcs e of $\mathcal{N}$ so that the following two conditions hold:

1. For each arc e, $0 \le f(e) \le c(e)$.

2. For each node v other than s and t,

$$\sum_{uv \in \mathcal{N}} f(uv) - \sum_{vw \in \mathcal{N}} f(vw) = 0.$$

The first condition ensures that the capacity of an arc is not exceeded. The second is called *conservation of flow* and says that the flow into an intermediate node equals the flow out of it. In a network flow problem, we want to send goods from node s to node t along the arcs of $\mathcal{N}$ and we wish to maximize the flow of goods. In terms of the flow function f, the *(total) flow* F in a network $\mathcal{N}$ can be measured at the sink t as the flow into t minus the flow out of t, so

$$F = \sum_{ut \in \mathcal{N}} f(ut) - \sum_{tw \in \mathcal{N}} f(tw).$$

A network is displayed in Figure 12.1.

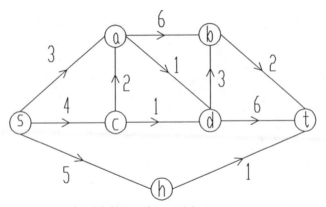

Figure 12.1   A network with integral capacities.

The problem of finding a maximum flow in a network was proposed and solved by Ford and Fulkerson [FF1]. They later wrote a whole book [FF2] on network flows. Their technique for finding a maximum flow in a network by successive calculations used semipaths from s to t that could increase the total flow. (Recall that in a semipath, one need not follow the direction on an arc.)

The *slack* $sl(e)$ of an arc $e \in \mathcal{N}$ is the amount by which its capacity exceeds its flow, that is, $s(e) = c(e) - f(e)$. If $s(e) = 0$, arc e is *saturated*. For a semipath P from s to t that traverses e in the correct direction, $s(e)$

measures the maximum amount of additional flow that e could tolerate. If $f(e) > 0$ and P traverses arc $e = uv$ in the wrong direction (that is, from v to u), then to increase the flow in $\mathcal{N}$, $f(e)$ would have to be decreased. The effect would be to divert some of the goods that flowed into uv and send them through some other arc leaving node u. An *augmenting path* in $\mathcal{N}$ is a semipath from s to t that can be used to increase the flow in $\mathcal{N}$. The *augmenting flow* for such a semipath P is the minimum of the values $s(e)$ for forward traversed arcs e and $f(e)$ for backward traversed arcs. Ford and Fulkerson [FF1] established the importance of augmenting paths with the following result.

Theorem 12.1   The flow F in a network $\mathcal{N}$ is maximum if and only if there is no augmenting path from s to t. $\square$

One shortcoming of the original Ford and Fulkerson algorithm for finding the maximum flow in a network was that in certain special cases it could be extremely inefficient. This difficulty was overcome when Edmonds and Karp [EK1] modified the algorithm to use a shortest possible augmenting path at each step. Thus they label semipaths using a breadth-first search. Virtually all maximum flow algorithms incorporate the concept of augmenting paths; it is the technique of selecting the paths that distinguishes them. We give an outline of a labeling procedure for finding a maximum flow based on the algorithm of Edmonds and Karp.

Algorithm 12.1   Maximum Flow (outline)
Procedure LABEL($\mathcal{N}, u, v, label$)
begin
    if traversed uv forward and $s(uv) > 0$ **then**
      $label(v) = uv$
    else if traversed uv backward and $f(vu) > 0$ **then**
      $label(v) = vu$ (thus a backward traversal)
end LABEL
Procedure INCREMENT($\mathcal{N}, label$)
begin
      backtrack to find augmenting path P
      for each v on P **do**
        if $label(v) = uv$ **then** $incr(v) = s(uv)$
        else if $label(v) = uv*$ **then** $incr(v) = f(vu)$
      $inc = \min\{incr(v) : v \text{ on P}\}$
      for each forward arc e in P, $f(e) = f(e) + inc$

```
        for each backward arc e in P, f(e) = f(e) − inc
        delete all node labels
end INCREMENT
begin                                                    (driver algorithm)
        for all arcs e do
            f(e) = 0
        while possible or until t is reached do
        begin
            LABEL additional nodes in BFS manner
            if t reached then
                Increment(N, P)
        end
end                                                                    □
```

Figure 12.2a gives the labeling obtained after completing the first pass through the while loop in Algorithm 12.1. The pair listed on each arc e is $c(e), f(e)$. Since node t has been labeled, we backtrack and find the path $P = sft$ and $inc = 1$. So $f(sf)$ and $f(ft)$ each get incremented by 1 and the labeling process starts again. After several more passes through the while loop, the network in Figure 12.2b is obtained. By checking the flow into t, we see that the maximum flow F for $\mathcal{N}$ is 7.

Since there are no additional augmenting paths in Figure 12.2b, Theorem 12.1 guarantees that the flow in $\mathcal{N}$ is maximum. Another theorem of Ford and Fulkerson [FF1] uses the concept of a cut to characterize the value of a maximum flow. A *cut* in $\mathcal{N}$ is a set of arcs that separate the source and the sink. Hence if M is a cut in $\mathcal{N}$, then any directed path from s to t must contain an arc of M. The *capacity $C(M)$ of a cut* M is the sum of the capacities of its arcs, that is, $C(M) = \sum_{e \in M} c(e)$.

Theorem 12.2 In any network $\mathcal{N}$, the value of a maximum flow equals the value of a minimum cut.   □

According to this theorem, there is a cut M in Figure 12.2b whose capacity is 7. The cut $M = \{sa, cd, fd, ft\}$ is a minumum cut for $\mathcal{N}$ and has capacity 7.

The efficiency of Algorithm 12.1 is $O(p^2 q)$. Other efficient techniques have been developed since the work of Edmonds and Karp. They generally use the concept of "layered networks." Thorough discussions of such algorithms are given in Gould [G10] and Sysło, Deo, and Kowalik [SDK1].

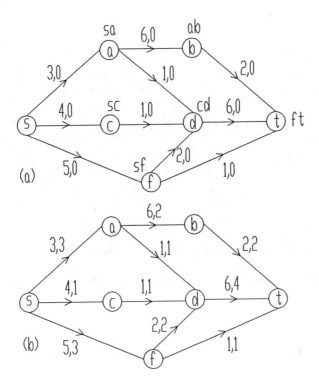

Figure 12.2   Generating a maximum flow in a network.

Matchings in Bipartite Graphs

We saw in §11.4 that finding a maximum matching in a graph can be rather complicated. However, when G is bipartite, we can simplify the process by using network flows. Suppose that G is bipartite with A and B as the two sets in the partition of $V(G)$. We obtain a maximum matching of G by the following algorithm.

Algorithm 12.2   Maximum Matching in a Bipartite Graph
begin
    input G, and sets A and B partitioning $V(G)$
    orient G by directing all edges if G from A to B
    add two new nodes s and t
    add arcs sa and bt for $a \in A$ and $b \in B$
    assign capacity 1 to each arc
    find a maximum flow in the resulting network using Algorithm 12.1
    maximum matching $M = \{ab : a \in A, b \in B, \text{ and } f(ab) = 1\}$
end   □

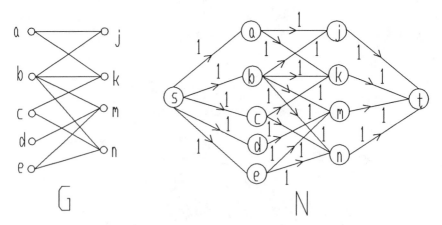

Figure 12.3   A bipartite graph G and its related network $\mathcal{N}$.

Figure 12.3 shows a bipartite graph and the network generated by Algorithm 12.2.

EXERCISES 12.1

1. Find a minimum cut for Figure 12.2b besides the one mentioned.
2. Every network has at least one flow and at least one maximum flow.
3. Find a maximum flow for each of the two networks in Figure 12.4.
4. If $\mathcal{N}$ is a network with every capacity integral, then $\mathcal{N}$ has a maximum flow where the flow on each arc is integral.
5. If $\mathcal{N}$ has no directed s-t path, then the maximum flow in $\mathcal{N}$ is zero.
6. Find a maximum flow for network $\mathcal{N}$ in Figure 12.3. Then give the corresponding maximum matching.
7. Use matchings in bipartite networks to prove that
    $\kappa(K_{m,n}) = \min\{m, n\}$.

12.2   MINIMUM SPANNING TREES

In §11.2 we saw two ways to generate spanning trees for a connected graph, namely, using either the BFS or DFS algorithm. An important related problem for networks and weighted graphs is to find a spanning tree for which the total weight is minimum.

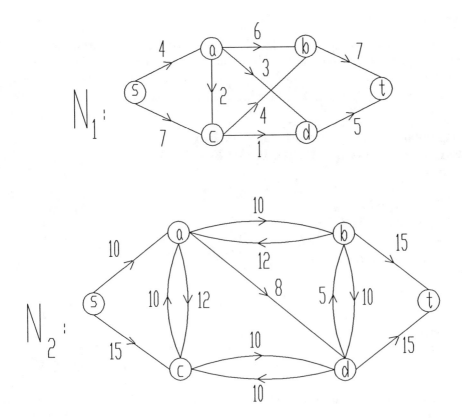

Figure 12.4   Two networks to flow through.

Weighted Graphs

We discussed weighted graphs briefly in §6.3 when considering distance matrix realizability problems. Each edge e in a weighted graph W has a weight $w(e)$ associated with it. Weighted graphs are also called undirected networks or simply networks when the context allows. We shall use this convention. The weight $w(e)$ might represent distance or travel time between the endnodes of e. The *length of a path* in W is the sum of the weights on its edges. The *distance $d(u,v)$ between u and v* is the length of a shortest path joining them.

Since distance is a metric for networks (at least when the edge weights represent true distances so that the triangle inequality holds), the concepts and algorithms described in earlier chapters for graphs have natural extensions. However, the algorithms must be altered to take the weights into account.

Kruskal's Algorithm

A *minimum spanning tree (MST)* is a spanning tree with minimum total weight. There are two standard algorithms for finding an MST in a graph. The first is a "greedy" method due to Kruskal [K7]. It continually selects a new edge of minimum weight which does not create a cycle.

Algorithm 12.3   Kruskal's Algorithm
Function Procedure REACHABLE(u, v, H)
begin
    do BFS in H with u as *root* to get tree T_u
    if $v \in T_u$ **then**
       $REACHABLE = true$
    else $REACHABLE = false$
end and **return** value of $REACHABLE$
begin
    input W in terms of ordered pairs $(uv, w(uv))$
    $T = \emptyset$
    $weight = 0$
    $tnodes = \emptyset$ (keeps track of the nodes of the tree)
    sort pairs $(uv, w(uv))$ onto stack, largest to smallest by weight
    $t = 0, \quad n = q$
    while $t < p - 1$ and $n > 0$ **do**
    begin
      remove next pair $uv, w(uv))$ from stack
      $n = n - 1$
      $first =$ one endnode of uv
      $second =$ other endnode of uv
      if $first \notin tnodes$ or $second \notin tnodes$ **then**
      begin
        $tnodes = tnodes \cup \{first, second\}$
        $t = t + 1$
        $T = T \cup \{uv\}$
        $weight = weight + w(uv)$
      end
      else if REACHABLE$(first, second, T)$ **then**
      begin
        $tnodes = tnodes \cup \{first, second\}$
        $t = t + 1$
        $T = T \cup \{uv\}$
        $weight = weight + w(uv)$

      end
    end
    if $t < p - 1$ **then**
      print "network is disconnected"
    else output T and *weight*
end   □

In Figure 12.5, we display successive networks in the process of generating an MST by using Kruskal's algorithm for a given network W.

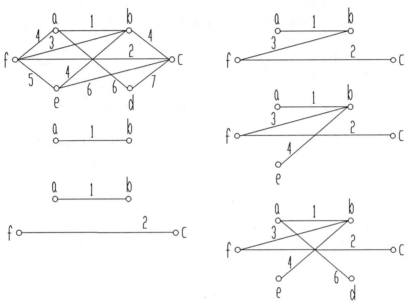

Figure 12.5   Generating an MST with Algorithm 12.5.

For a (p, q)-graph, the sorting of the pairs can be done in $O(q \lg q)$. Since the rest of the algorithm independently runs through at most the q pairs, the efficiency of Algorithm 12.3 is $O(q \lg q)$. We've chosen this version of the algorithm for simplicity. One could improve the efficiency to $O(q + p \lg q)$ by using a "heap" rather than completely presorting the list of pairs. See [SDK1,p.255] for that approach.

Note that Algorithm 12.3 discovers whether W is disconnected and could easily be modified to output a minimum spanning forest in that case.

Prim's Algorithm

A second standard technique for finding an MST is due to Prim [P9], and operates by continually trying to attach a new edge of minimum weight to the existing tree. Thus, unlike Kruskal's algorithm, the graph at each intermediate step of Prim's algorithm is connected.

Algorithm 12.4   Prim's Algorithm
begin
    input W in terms of ordered pairs $(uv, w(uv))$ and adjacency lists
    for some v **do**   (initialize the tree at an arbitrary node)
      if $N(v) \neq \emptyset$ **then**
        select $u \in N(v)$ with $w(uv)$ minimum
        $tnodes = \{v, u\}$
        $T = \{uv\}$
        $weight = w(uv)$
      end
    while some $x \notin tnodes$ and $(x \in N(v)$ for $v \in tnodes)$ **do**
    begin
      select $x \notin tnodes$ with $w(ux) = \min\{w(vx) : v \in tnodes, x \in N(v)\}$
      $T = T \cup \{ux\}$
      $tnodes = tnodes \cup \{x\}$
      $weight = weight + w(ux)$
    end
    if $tnodes \neq V$ **then**
      print "W is disconnected"
    else
      output T and $weight$
end   □

The efficiency of Prim's algorithm is $O(p^2)$. Thus, it would be preferred over Kruskal's algorithm for large dense graphs, while Kruskal's algorithm is generally better when W is sparse. Other more efficient algorithms may use a "parallel processing" approach (with special data structures) for generating an MST. This approach begins at some collection of nodes and "grows" small minimum weight trees around those nodes and then attempts to "patch them together" to form an MST.

We have presented two polynomial algorithms for finding an MST. It is interesting to note that there is no known polynomial algorithm for the corresponding problem for digraphs. However, if we expand the problem slightly for digraphs, the situation is better. A *branching* of a digraph is an acyclic subdigraph for which $id(v) \leq 1$ for each node v. There are polynomial algorithms for finding a maximum (or minimum) weight branching of a digraph. Gibbons [G5] and Gould [G10] discuss algorithms for branchings.

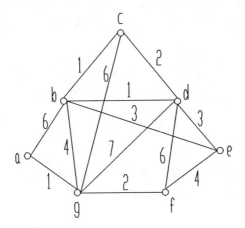

Figure 12.6   A network to span through.

EXERCISES 12.2

1. Find an MST for the network in Figure 12.6 using Kruskal's algorithm.

2. Use Prim's algorithm in Exercise 1.

3. Construct a branching which is not a spanning tree for a digraph.

4. State conditions under which a branching of a digraph becomes a directed spanning tree.

5. Suppose that T is an MST for a network W and $e \in W - T$. Then $T + e$ has one cycle C and $w(e) \geq w(e')$ for each $e' \in C$.

6. Prove that a spanning tree found by Algorithm 12.5 does indeed have minimum weight.

7. If $w(e)$ is distinct for each edge e, then Kruskal's Algorithm and Prim's Algorithm each generate the same unique tree.

12.3   TRAVELING SALESMAN PROBLEM

A salesman plans to visit various cities to show his merchandise. He would like to stop in each city once and return to the home office while minimizing the total distance traveled. This is called the *traveling salesman problem*. Only certain pairs of cities have direct routes joining them. Let weight $w(uv)$ represent the direct route distance between cities u and v. We must then find a minimum-weight hamiltonian cycle in the resulting network. This problem is certainly $\mathcal{NP}$-complete, because if all edge weights are made to equal one, the problem reduces to the $\mathcal{NP}$-complete problem of finding a hamiltonian cycle in a graph.

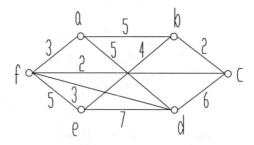

Figure 12.7   A network for the traveling salesman's cities.

The network in Figure 12.7 has exactly 6 hamiltonian cycles:

1. a, b, c, d, e, f, a has weight 28;
2. a, b, c, f, e, d, a has weight 26;
3. a, b, e, d, c, f, a has weight 27;
4. a, b, e, f, c, d, a has weight 27;
5. a, d, c, b, e, f, a has weight 25; and
6. a, d, e, b, c, f, a has weight 23;

Thus, we choose a, d, e, b, c, f, a as a minimum-weight hamiltonian cycle.

We should mention that there are various versions of the traveling salesman problem. Sometimes the overriding concern is minimizing total cost rather than distance, even if this requires revisiting a node. In this situation, the triangle inequality rarely holds. However, here we shall focus on the problem in which a node cannot be revisited and the triangle inequality does hold.

The traveling salesman problem involves greater difficulty than finding a hamiltonian cycle. If a network has many hamiltonian cycles, we must find one with minimum total weight. The complete graph K_p has $(p-1)!/2$ hamiltonian cycles. Even for moderate values of p, checking all such cycles would be ludicrous. For example, it might require a century of computer time to check K_{20} in this manner. Thus other approaches are called for. Two standard approaches are approximation techniques and branch-and-bound methods.

Approximation Techniques

In a graph having a hamiltonian cycle, we want an approximate algorithm that will produce a hamiltonian cycle C in polynomial time so that the total distance traveled $w(C)$ is reasonably close to an optimal solution most of the time. Of course, one might first want to decide what "reasonably close" and "most of the time" mean. But there is another difficulty: how can we tell if we are close to an optimal solution without knowing an optimal solution? We first find a lower bound for the minimum total distance $w(C)$.

Theorem 12.3   If T is a minimum weight spanning tree for network W with a hamiltonian cycle and e is a minimum weight edge in $W - T$, then the minimum total distance of a hamiltonian cycle is at least $w(T) + w(e)$.

Proof   Let C be a hamiltonian cycle with minimum total distance for W. For each edge $e' \in C$, $C - e'$ is a spanning tree of W and $w(T) \le w(C - e')$. Furthermore, at least one such edge e' is in $W - T$, and a minimum weight edge e in $W - T$ satisfies $w(e) \le w(e')$. Hence,
$$w(T) + w(e) \le w(C - e') + w(e') = w(C). \qquad \square$$

If there is not much variation in the weights on the edges of W, the lower bound $w(T) + w(e)$ of Theorem 12.3 is usually a good approximation for the optimal distance.

Common approximation techniques for the traveling salesman problem use a local greedy approach. Although it is not necessary, these often use an implicit assumption that the graph is complete, with $w(uv)$ representing travel cost. Also the triangle inequality usually is assumed. The *nearest neighbor method* begins with an arbitrary node (a trivial path) and extends it and subsequent paths with a new node nearest to an endnode of the path. When all nodes have been visited, the path is closed by adding the edge joining the endnodes to form a cycle. This approach produces early savings in edge weights, but could be far from optimal when the cycle is formed.

Another common technique is the *insertion method*, which generates an initial cycle and continually expands it to a larger cycle by adding a new node adjacent to a pair of consecutive nodes on the cycle at minimum additional cost. Again, a complete graph is usually assumed here. In our description of the algorithm a procedure is used to find the next node z to insert and where to insert it.

Algorithm 12.5   Traveling Salesman Insertion Method
Procedure ADDNODE($W, C, n, total$)
begin
    for each edge uv on C **do**
      find $x \in W - C$ minimizing $t(x, uv) = w(ux) + w(xv) - w(uv)$
      let z be node for which $t(z, uv)$ is minimum
      $C = C - \{uv\} \cup \{uz, zv\}$
      $total = total - t(z, uv)$
      $n = n + 1$
end ADDNODE

begin
    input p and weighted complete graph W
    let C be minimum weight edge uv (initial "cycle")
    find node x so that $w(ux) + w(xv)$ is minimum
    $C = u, x, v, u$
    $n = 3$ (keeps track of cycle size)
    $total = w(uv)$
    while $n < p$ **do**
      ADDNODE($W, C, n, total$)
    output C and $total$
end   □

The insertion method is much more effective than the nearest neighbor method, and produces a cycle having at most twice the optimal length in the worst case.

Other algorithms to get approximate solutions to the traveling salesman problem involve finding an initial hamiltonian cycle and then performing edge exchanges to improve the total weight. Suppose the cycle is $C = v_1 v_2 \cdots v_p v_1$. In a 2-exchange, we look for a pair of edges $v_i v_{i+1}$ and $v_j v_{j+1}$ such that $w(v_i v_j) + w(v_{i+1} v_{j+1}) < w(v_i v_{i+1}) + w(v_j v_{j+1})$. If such a pair is found, edges $v_i v_{i+1}$ and $v_j v_{j+1}$ of C are replaced by $v_i v_j$ and $v_{i+1} v_{j+1}$. Algorithms based on 2-exchanges and 3-exchanges together with computer implementations are given in [SDK1].

Branch-and-Bound Method

Branch-and-bound is an optimization technique that limits the amount of searching required to find an optimal solution to a problem. For the traveling salesman problem, a rooted binary (search) tree is generated in which each branch at a node is determined by whether a particular edge is to be included in the cycle. We then obtain lower bounds on the total weight of a cycle that would eventually be found via each given branch of the search tree. Thus as we travels down through the tree, increasing lower bounds are obtained. When we find a hamiltonian cycle C and determine the total distance traveled, we can eliminate all branches of the search tree whose weight bounds exceed $w(C)$. We only continue to branch and bound through the remaining branches looking for additional cycles.

The branch-and-bound method will always lead to an optimal solution. Although it does save much unnecessary searching, it is not a panacea — the traveling salesman problem will still require an exponential amount of time in the worst case. A detailed discussion of this technique for the traveling salesman problem is given in Reingold, Nievergelt, and Deo [RND1] as well as in [SDK1] which gives a computer implementation of the algorithm.

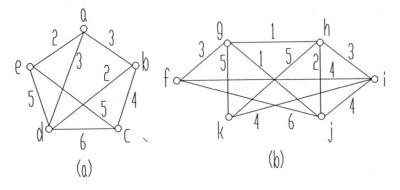

Figure 12.8   Two networks to travel through.

EXERCISES 12.3

1. Use Theorem 12.3 and Kruskal's Algorithm to obtain a lower bound on the total distance for the traveling salesman problem on Figure 12.8a.

2. Use Prim's Algorithm and Figure 12.8b in Exercise 1.

3. Solve the traveling salesman problem for the two networks in Figure 12.8.

4. If the weights on edges do not satisfy the triangle inequality, then the nearest neighbor method could be arbitrarily bad.

5. Determine the efficiency of Algorithm 12.5

12.4   SHORTEST PATHS

We discussed path algorithms for graphs in §11.2. The two common techniques, breadth first search and depth first search must be modified to deal with networks. We first consider the problem of finding a shortest path between two nodes in a network.

Dijkstra's Algorithm

Each arc of a digraph is given a nonnegative weight representing the distance from one of its endnodes to the other. We want to find the distance between two specified nodes of this network. The algorithm due to Dijkstra [D3] is reminiscent of BFS and Prim's Algorithm; the difference is that a node v receives a temporary label $t(v)$ representing its distance from the root r. The label is updated each time a shorter path from r is found. The input to the algorithm is the list of nodes, their adjacency lists, arc weights, and specification of the root and the target node. Let $N(v)$ denote the adjacency list of node v, that is, the list of nodes adjacent *from* v. Of course, this algorithm applies to undirected networks as well as directed ones; in the latter case $N(v)$ has been called the outneighborhood of v.

Algorithm 12.6   Dijkstra's Algorithm
begin
    input W, *root*, *target*
    $t(root) = 0$
    for $v \neq root$ **do**
      $t(v) = \infty$
    $T = V(W)$
    $u = root$
    while $u \neq target$ **do**
    begin
      for each $v \in N(u)$ **do**
        if $v \in T$ and $t(v) > t(u) + w(uv)$ **then**
          $t(v) = t(u) + w(uv)$
      $T = T - \{u\}$
      let u be node in T for which $t(u)$ is minimum
    end
    output $t(target)$
end   □

The complexity of Dijkstra's Algorithm on a (p, q)-graph is $O(q)$ which equals $O(p^2)$, so like BFS is quite an efficient method for finding the distance. It can be easily modified to find the distance from the root to all the other nodes in the graph still with complexity $O(p^2)$.

By applying the modified form of Dijkstra's Algorithm with each node successively serving as root, one can determine the distance from each node to each other node. The resulting algorithm has complexity $O(pq) = O(p^3)$. Note that it is also quite easy to modify Algorithm 12.6 with a labeling scheme to recover a shortest path (rather than just its length) from the root to each node without increasing the complexity. In this way we can generate a spanning tree where the distance from each node to the root in the tree is the same as in the original network. This is analogous to the situation for BFS for graphs.

Floyd's Algorithm

Although true distances are always nonnegative, the weights in a network need not be. The one shortcoming of Dijkstra's Algorithm is that it can fail if the arc weights are allowed to be negative. A number of algorithms have been developed to handle this case, the most popular of which is due to Floyd [F5]. This $O(p^3)$ algorithm finds the distance between all pairs of nodes in a network which may have negative arc weights but has no negative cycle (a cycle for which the sum of its arc weights is negative). Negative cycles are not allowed because by continuously traversing such a cycle the path length has no lower bound.

The nodes of W are $v_1, v_2, \ldots, v_p$ and each node v_j receives a temporary labels $t_k(v_j, v_j)$ representing the distance from v_i to v_j using only nodes in $\{v_1, v_2, \ldots, v_k\}$. The labels are updated each time a shorter path is found. The input to the algorithm is W in the form of a list of nodes, their adjacency lists, and arc weights. Let $N(v)$ denote the adjacency list of node v, that is, the list of nodes adjacent *from* v.

Algorithm 12.7   Floyd's Algorithm
```
begin
    input W
    for i = 1 to p do
        for j = 1 to p do
            if v_j ∈ N(v_i) then
                t_0(v_i, v_j) = w(v_i v_j)
            else t_0(v_i, v_j) = ∞
    for k = 1 to p do
```

for $i = 1$ to p **do**
    for $j = 1$ to p **do**
        $t_k(v_i, v_j) = \min\{t_{k-1}(v_i, v_j), t_{k-1}(v_i, v_k) + t_{k-1}(v_k, v_j)\}$
end   □

Algorithm 12.7 is extremely easy to implement by computer. In fact, the computer code differs only slightly from Algorithm 12.7. Note also that the algorithm works equally well for undirected networks as for directed ones. Furthermore, a labeling scheme could be added to recover each shortest path rather than just its length.

We have described the two most popular path algorithms for networks. Of course, there are other algorithms for this and related problems. An extensive bibliography of such algorithms is given in Pierce [P2]. We shall now consider an optimization problem where it is necessary to find shortest paths between certain pairs of nodes.

Chinese Postman Problem

Suppose a postman wishes to begin at the post office and deliver mail along each street in his route and return to post office while traversing each street exactly once. We should recognize that this is possible if and only if the graph that models his route is eulerian. From §4.3, we recall that the graph is eulerian if and only if each node has even degree. Of course, not all delivery routes will have this property. Thus the condition is relaxed to insure that each street is traversed at least once.

The *Chinese postman problem* is to find a minimum length closed spanning walk that includes each edge of a network W at least once. This problem was named this way because it was first described in the context of the postman 30 years ago in a paper by Guan [G11]. Several years ago in New York, Guan gave a survey talk about this problem and began, "I am Chinese, but I am not a postman." (Incidently, he is, in fact, the president of a university in China.) If the total weight of W is k and $w(e) \geq 0$ for each edge e, then the total distance the postman must travel is between k and $2k$. The worst case occurs when W is a tree. If this case each edge must be traversed twice. At the other extreme are the eulerian graphs for which we need to use each edge only once and accumulate total distance k.

Perhaps, the best known algorithm to generate an eulerian trail in an eulerian graph is Fleury's algorithm. The key to the algorithm is to avoid using a bridge unless there is no other alternative. The one difficulty with the method is that one must repeatedly determine whether a given edge is a bridge. Of course, this can be done fairly easily as follows: to check whether uv is a bridge in G, do a BFS or DFS in $G - uv$ using u as root. If v is reached, uv is not a bridge, otherwise it is. Since the weights are unimportant at this point, we state the algorithm for graphs.

Algorithm 12.8   Fleury's Algorithm
begin
    input G, adjacency lists, and q
    for some $v \in G$ **do**
    begin
        $C = v$ (initial node of eulerian walk)
        $x = v$
    end
    $m = 0$ (counts number of edges used)
    while $m < q$ **do**
    begin
      if $|N(x)| > 1$ **then**
        find $u \in N(x)$ such that xu is not a bridge of G
      else
        let u be unique node in $N(x)$
      $G = G - xu$
      $N(x) = N(x) - \{u\}$
      $N(u) = N(u) - \{x\}$
      $m = m + 1$
      $C = C, u$ (insert u as next node of walk C)
      $x = u$
    end
    output C
end   □

Now that we have an algorithm to find an eulerian trail, we can proceed with the Chinese postman problem. Goodman and Hedetniemi [GH1] observed that the Chinese postman problem is solved by finding the minimum number of edges that must be duplicated to produce a weighted eulerian multigraph. Since a multigraph is also eulerian if and only if all its degrees are even, we must find a set of minimum length paths joining pair of odd degree nodes. If the total length of the paths added to the graph is minimum, then an eulerian trail in the resulting weighted multigraph describes a solution to the Chinese postman problem.

Given a network W with all weights nonnegative, let W' be the weighted complete graph whose nodes are the nodes of odd degree in W such that each edge $e = uv$ of W' has weight equal to the distance

between u and v in W. Let $W''(S)$ be the weighted multigraph formed
from W by adding in the edges of the set S of given shortest paths joining
certain pairs of odd degree nodes of W.

Algorithm 12.9   Chinese Postman Algorithm
begin
    input W
    determine set T of nodes of odd degree in W
    find the distance between each pair of nodes of T
    form W'
    find a minimum weight perfect matching M of W'
    identify the set S of edges in W corresponding to M
    form W''
    use Algorithm 12.8 to find an eulerian trail of W''
end   □

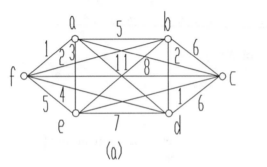
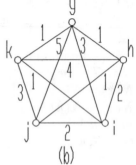

Figure 12.9   More networks to travel through.

EXERCISES 12.4

1. Use Dijkstra's Algorithm to find the length of a shortest path
   between a and e in Figure 12.9a.
2. Use Floyd's Algorithm and Figure 12.9b in Exercise 1.
3. Modify Dijkstra's Algorithm to find shortest paths between *all* pairs
   of nodes in a network with no negative edges.
4. Apply Fleury's Algorithm to the network of Figure 12.9a.
5. An old method of Hierholzer [H23] for finding an eulerian trail in
   an eulerian graph G finds pairwise-edge-disjoint cycles partitioning
   $E(G)$. It then finds nodes they have in common to patch the cycles

together to form an eulerian trail. Describe this algorithm in coding similar to that we have been using.

6. Solve the Chinese Postman problem for the network in Figure 12.9b.

7. Give an example of a network which has some negatively weighted arcs (but no negative cycles) for which Dijkstra's Algorithm does not work.

12.5   CENTERS

In Chapter 2 we studied centrality concepts, and algorithms to find several types of centers were discussed in §11.3.

The definitions of distance, center and median of a network are analogous to those for a graph. A fundamental result for the center of graphs, which is the basis of many algorithm and theorems about centers also holds for networks.

Theorem 12.4   The center of any network lies in a single block.   □

We should mention at this point that in some applications, one also places weights on the nodes and obtains a *doubly-weighted network*. The *weight $\omega(v)$ at a node* could represent a demand for service. The *weighted distance $wd(u,v)$* from u to v is the product $\omega(v)d(u,v)$, where $d(u,v)$ is the minimum sum of the edge weights $w(e)$ for paths joining u and v. A doubly-weighted network is displayed in Figure 12.10b.

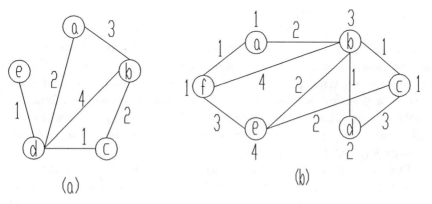

(a)   (b)

Figure 12.10   A network and a doubly-weighted network.

A motivation for doubly-weighted networks described in [B27] comes from facility location problems. The idea is that a densely populated neighborhood requires more service than a sparsely populated one, and its demand is roughly proportional to its population. Thus a weight is placed on a node to indicate the population for the corresponding neighborhood. Unfortunately, the resulting distance function is not a metric, so we can not take advantage of convexity properties of metrics for doubly-weighted networks. Facility location problems for doubly-weighted *trees* are studied in Berman, Simchi-Levi, and Tamir [BST1].

With the aid of Dijkstra's Algorithm (12.6) or Floyd's Algorithm (12.7), finding the center or median of a network becomes a simple task.

Algorithm 12.10   Center of Network
begin
    input W
    use Algorithm 12.7 to generate the distance matrix
    determine eccentricity of each node
    determine the center
end   □

Absolute Centers and Medians

In Chapter 2 we described several applications where centrality plays a key role. Two such examples were in emergency facility and service facility location problems. We generally want to locate an emergency facility at a central node of the graph to minimize response time to the farthest node. When a rural region is modeled by a network, the edge weights tend to be rather large. In this situation, one might not insist that an emergency facility be located at a node (intersection) if locating elsewhere would decrease the maximum response time. A *position* of a network W is either a node or an internal point on an edge of W. An *absolute central position* of W is a position whose distance to a farthest node is minimum. The *absolute center* of W is the set of all absolute central positions. Note that a bicentered tree has only one absolute center, namely, the midpoint of the edge joining the two centers.

The *absolute median* of a network is defined in an analogous way. Hakimi [H2] obtained the first algorithms for finding absolute medians and absolute centers of networks. There is also a large literature on these topics in a wide variety of journals. Christofides [C10] and Handler and Mirchandani [HM1] provide additional details on absolute centers or absolute medians.

EXERCISE 12.5

1. For the network in Figure 12.10a, determine

   a. the distance matrix.

   b. the center.

   c. the median.

2. Describe an algorithm to find the center of a doubly-weighted network.

3. Determine the distance matrix of the doubly-weighted network in Figure 12.10b.

4. The *branch weight* of a node v in a weighted tree T is the maximum sum of the weights in the branches of $T'- v$. Give an algorithm to find the branch weight of each node in a weighted tree.

12.6   CRITICAL PATH METHOD

Every large project consists of many activities. The manager of such a project must estimates of the duration of these activities and define precedence relations between them. For example, on a construction project, the electrical wiring must be installed before the wallboard, the walls do not get painted until the wallboard is installed, etc. Such problems are modeled using a network.

An *activity digraph* consists of an acyclic network $\mathcal{N}$. Each node of $\mathcal{N}$ represents an activity; the source is a node called *start* and the sink is called *finish*. The presence of arc uv indicates that activity u is an *immediate predecessor* of v, and v is a *successor* of u. Node *start* has no immediate predecessors and *finish* has no successors. The weight on arc uv is the time required to complete activity u. To begin an activity, the activities of all its immediate predecessors must have been completed.

For a given activity digraph, we would like to determine the minimum amount of time needed to complete all activities. A longest path in an activity digraph is called a *critical path*.

Theorem 12.5   The length of a critical path in an activity digraph equals the minimum time needed to complete the project.

Proof   Since each activity in a critical path must be completed before its successor is begun, the project completion time is at least the length of a critical path. The earliest possible start time for each activity v is the length of a longest path from *start* to *v*. Thus the earliest start time of *finish* is the length of a critical path. But when we begin *finish*, the project is complete. Thus the project can be completed in time equal to the length of a critical path.   □

The *critical path method* is a technique for finding an optimal scheduling of activities so that a project can be completed in the minimum time. We model the project with an activity digraph and start each activity at its earliest possible start time. In Table 12.1 we list a sequence of possible activities involved in the opening of a diner.

Table 12.1 OPENING A DINER

Activity	Description	Immediate Predecessors	Duration in Days
start	—	—	0
a	meet lawyer	*start*	3
b	meet accountant	*a*	2
c	negotiate lease	*a*	4
d	hook up electricity	*b, c*	1
e	issue stock	*b, c*	3
f	hook up gas	*b, c*	1
g	get state tax number	*b*	1
h	get license	*g*	1
i	set up equipment	*d, e, f*	2
j	city health inspection	*i*	1
finish	—	*h, j*	0

The activity digraph for Table 12.1 is given in Figure 12.11. For each activity *v*, the following important quantities are calculated to obtain an optimal schedule.

1. The earliest start time of *v* equals the maximum of the earliest finish times of its immediate predecessors.

2. The earliest finish time of *v* equals the sum of its earliest start time and its duration.

3. The latest start time of *v* equals its latest finish time minus its duration.

4. The latest finish time of *v* equals the minimum of the latest start times of its successors.

5. The slack time of *v* equals its latest start time minus its earliest start time.

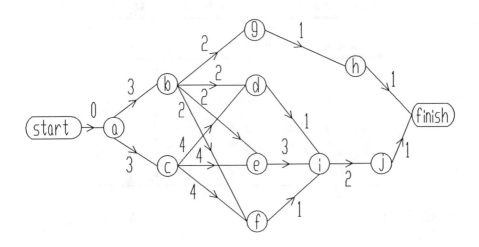

Figure 12.11   Activity digraph for opening the diner.

These quantities for the activities of Table 12.1 are given in Table 12.2. Since the earliest finish time of *finish* is the length of a critical path, the project of opening the diner can be completed in 13 days. Note that this algorithm amounts to finding a longest path in an acyclic digraph (similar to Dykstra's Algorithm for shortest paths).

Table 12.2 SCHEDULE TIMES FOR THE DINER

	start	*a*	*b*	*c*	*d*	*e*	*f*	*g*	*h*	*i*	*j*	*finish*
Duration	0	3	2	4	1	3	1	1	1	2	1	0
earliest start time	0	0	3	3	7	7	7	5	6	10	12	13
earliest finish time	0	3	5	7	8	10	8	6	7	12	13	13
latest start time	0	0	5	3	9	7	9	11	12	10	12	13
latest finish time	0	3	7	7	10	10	10	12	13	12	13	13
slack time	0	0	2	0	2	0	2	6	6	0	0	0

An extensive treatment of the critical path method, its computer implementation, and related scheduling problems are presented in [SDK1]. Chachra, Ghare, and Moore [CGM1] contains an abundant collection of genuine critical path method examples obtained from various corporations.

Table 12.3 A TABLE FOR EXERCISE 2

Activity	Immediate Predecessors	Duration
start	—	0
a	*start*	2
b	*start*	3
c	*a*	5
d	*a, b*	4
e	*c, d*	6
f	*b, c*	2
finish	*e, f*	0

EXERCISES 12.6

1. An activity is *critical* if its slack time is zero. Every activity on a critical path is critical.

2. For the data in Table 12.3

    a. Construct the activity digraph.

    b. For each activity v, compute all the quantities in Table 12.2

    c. Find a critical path and the project completion time.

3. For the activity digraph in Figure 12.12, perform steps b and c of Exercise 2.

4. An investment company prepares an annual report for its customers. This report will be mailed to all customers, stockholders, and selected individuals and organizations. The activities to be performed, their immediate predecessors and durations are shown in Table 12.4.

    a. What is the fewest number of weeks in which the report can be prepared?

    b. What are the critical activities?

    c. Find a critical path.

5. Explain why an activity digraph cannot contain a directed cycle or a transitive triple (a set of three arcs uv, vw, uw).

Table 12.4 PREPARING THE ANNUAL REPORT

Activity	Description	Immediate Predecessors	Duration in Weeks
start	—	—	0
a	decide on theme	*start*	1
b	write articles	a	4
c	do art work	a	2
d	layout report	b, c	2
e	prepare mailing list	d	1
f	proofread first draft	d	3
g	make final changes	f	2
h	check printed report	e, g	1
i	mail report	h	2
finish	—	i	0

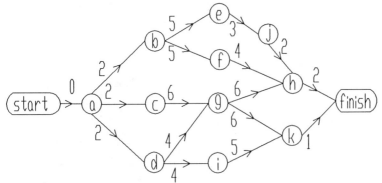

Figure 12.12   Activity digraph for Exercise 3.

Appendix

	q=0	q=1	q=2	q=3	q=4	q=5	q=6
p=1							
p=2							
p=3							
p=4							

p=5

q=0	q=1	q=2	q=3	q=4

q=5	q=6

p=5 (continued)

q=7	q=8	q=9	q=10

p=6

q=0	q=1	q=2	q=3

q=4

p=6 (continued)

q=5

q=6

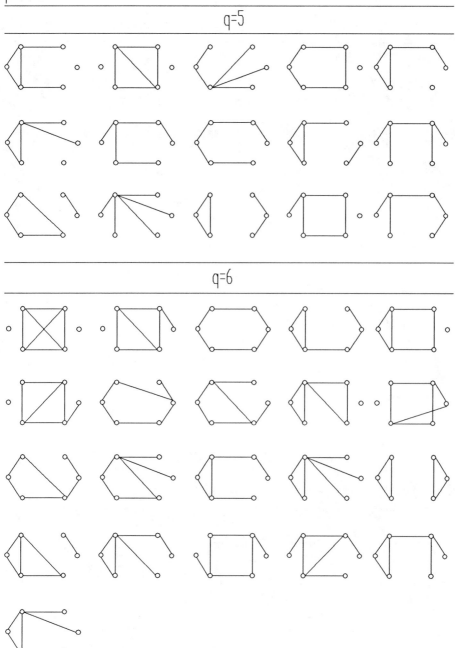

p=6 (continued)

q=7

q=8

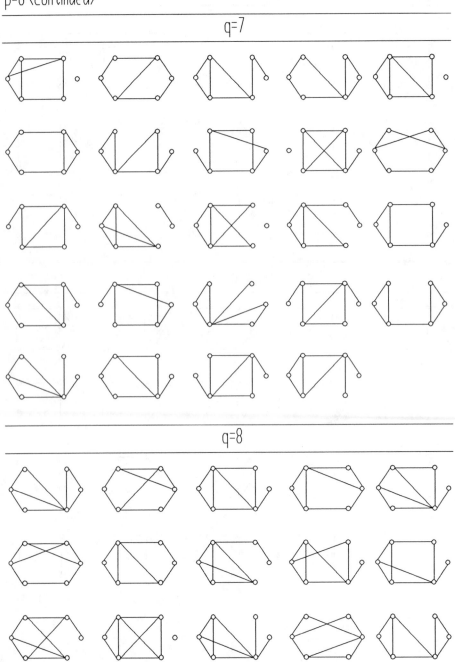

p=6 (continued)

q=8 (continued)

q=9

q=10

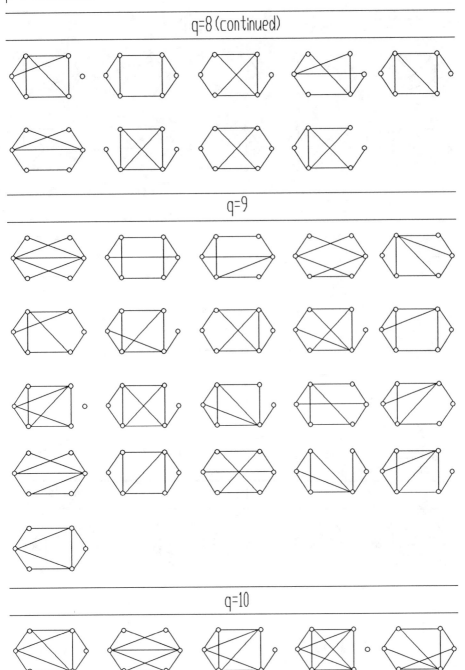

p=6 (continued)

q=10 (continued)

q=11

q=12

q=13	q=14	q=15

References

Adel'son-Velski, G.M.

[A1]   Example of a graph without transitive automorphism group. *Soviet Math. Dokl.* **10** (1969) 440-441.   [169]

Akiyama, J.

[AAA1]   (with K. Ando and D. Avis) Miscellaneous properties of equieccentric graphs. in *Convexity and Graph Theory, Proc. Conf. in Haifa, Israel, 1981*, North-Holland, Amsterdam (1984) 13-23.
                                                                        [39,41]

[AAA2]   (with K. Ando and D. Avis) Eccentric graphs. *Discrete Math.* **56** (1985) 1-6.   [99]

[AH1]   (with F. Harary) A graph and its complement with specified properties III: Girth and circumference. *Internl. J. Math. Math. Sci.* **2** (1979) 685-692.   [25,115]

Alagar, V.S.

[AOS1]   (with J. Opatrny and N. Srinivasan) Bigeodetic graphs. *Graphs and Combin.* **4** (1988) 379-392.   [204]

[AS1]   (with N. Srinivasan) Criticality concepts in geodetic blocks. *J. Math. Phys. Sci* **22** (1988) 241-250.   [148]

Alavi, Y.

[AW1]   (with J.E. Williamson) Panconnected graphs. *Studia Sci. Math. Hungar.* **10** (1975) 19-22.   [93]

Albertson, M.O.

[A2]   Finding hamiltonian cycles in Ore graphs. *Congressus Numer.* **58** (1987) 25-27.   [252]

            see also p. 186

Amin, A.T.

[ASS1]   (with K.T. Siegrist and P.J. Slater) Pair-connected reliability of a tree and its distance degree sequences. *Congressus Numer.* **58** (1987) 29-42.   [187]

Ando, K.   see [AAA1], and [AAA2]

Aschbacher, M.

[A3]   The non-existence of rank three permutation groups of degree 3250 and subdegree 57. *J. Algebra* **19** (1971) 538-540.   [101]

290

Avis, D.   see [AAA1], and [AAA2]

Ayel, J.

[A4]   Longest paths in bipartite digraphs. *Discrete Math.* **40** (1982) 115-
            118.   [221]

Babai, L.

[B1]   On the abstract group of automorphisms. *Combinatorics,* LMS
            Lecture Notes **52** (1981) 1-40.   [161]

Balaban, A.T.   see p. 196

Bandelt, H.-J.

[BM1]   (with H.M. Mulder) Distance-hereditary graphs. *J. Combin. The-
            ory* **41B** (1986) 182-208.   [152–154]

Barefoot, C.A.

[BCDEF1]   (with L.H. Clark, J. Douthett, R.C. Entringer, and M.R. Fellows)
            Cycles of length 0 modulo 3 in graphs. *Graph Theory, Combina-
            torics, and Applications,* Wiley, New York (to appear).   [114]

Bauer, D.

[B2]   Line-graphical degree sequences. *J. Graph Theory* **4** (1980) 219-
            232.   [202]

[BBB1]   (with G.S. Bloom and F.T. Boesch) Edge to point degree lists and
            extremal triangle-free point degree regular graphs. *Proceedings in
            Graph Theory (Waterloo, 1982),* Academic Press, Toronto (1984)
            93-104.   [202]

[BS1]   (with E.F. Schmeichel) Long cycles in tough graphs. *Research Re-
            ports in Math.* **8612** Stevens Institute of Technology (1986). [113]

[BT1]   (with R. Tindell) Graphs with prescribed connectivity and line
            graph connectivity. *J. Graph Theory* **3** (1979) 393-395.   [63]

[BMSV1]   (with A. Morgana, E.F. Schmeichel, and H.J. Veldman) Long cy-
            cles in graphs with large degree sums. *Discrete Math.* (to appear).
                                                                          [113]

[BSV1]   (with E.F. Schmeichel and H.J. Veldman) Some recent results on
            long cycles in tough graphs. *Graph Theory, Combinatorics, and
            Applications,* Wiley, New York (to appear).   [113]

Behzad, M.

[BS2]   (with J.E. Simpson) Eccentric sequences and eccentric sets
            in graphs. *Discrete Math.* **16** (1976) 187-193.   [179]

Beineke, L.W.

[B3]   Derived graphs and digraphs. *Beiträge zur Graphentheorie*, Teub-
            ner, Leipzig (1968) 17-33.   [28]

[BW1]   (with R.J. Wilson) *Selected Topics in Graph Theory 3*. Academic
            Press, San Diego (1988).   [94]

Berge, C.

[B4]   Two theorems in graph theory. *Proc. Nat. Acad. Sci. U.S.A.* **43**
            (1957) 842-844.   [243]

[B5]   *Graphs*. 2nd revised edition, North-Holland, Amsterdam (1985).
                                                                          [61]

Berman, D.M.   see p. 186

Berman, O.

[BST1]   (with D. Simchi-Levi and A. Tamir) The minimax multistop loca-
            tion problem on a tree. *Networks* **18** (1988) 39-49.   [276]

Bermond, J.-C.

[BB1]   (with B. Bollobás) The diameter of graphs - a survey. *Congressus
            Numer.* **32** (1981) 3-27.   [38,105]

[BBPP1]   (with J. Bond, M. Paoli, and C. Peyrat) Graphs and interconnec-
            tion networks: diameter and vulnerability. *Surveys in Combina-
            torics*, London Math. Soc. Lecture Notes **82** (1983) 1-30. [62]

[BGHS1]   (with A. Germa, M.C. Heydemann, and D. Sotteau) Longest paths
            in digraphs. *Combinatorica* **1** (1981) 337-341.   [221]

Bhattacharya, D.

[B6]   The minimum order of *n*-connected *n*-regular graphs with specified
            diameters. *IEEE Trans. Circuits Syst.* **CAS32** (1985) 407-409.
                                                                     [108–109]

Bielak, H.

[B7]   Minimal realizations of graphs as central subgraphs. *Graphs, Hy-
            pergraphs, and Matroids* (Zagań, 1985), Higher College Engnrg.,
            Zielona Góra (1985) 13-24.   [41]

[BS3]   (with M.M. Sysło) Peripheral vertices in graphs. *Studia Sci. Math.
            Hungar.* **18** (1983) 269-275.   [57]

Biggs, N.L.

[B8]   *Algebraic Graph Theory*. Cambridge University Press, London
            (1974).   [101,133,167,169–170]

[BBS1]   (with A.G. Boshier and J. Shawe-Taylor) Cubic distance-regular graphs. *J. London Math. Soc.* **33** (1986) 385-394.   [170]

[BS4]   (with D.H. Smith) On trivalent graphs. *Bull. London Math. Soc.* **3** (1971) 155-158.   [167]

Birkhoff, G.D.

[B9]   *Lattice Theory*. American Mathematical Society, Providence (1967).   [75]

Bloom, G.S.

[BKQ1]   (with J.W. Kennedy and L.V. Quintas) Distance degree regular graphs. *Theory and Applications of Graphs*, Wiley, New York (1981) 95-108.   [45,181,185]

[BKQ2]   (with J.W. Kennedy and L.V. Quintas) Some problems concerning distance and path degree sequences. *Lecture Notes in Math.* **1018** (1983) 179-190.   [182,185,204]

           see also [BBB1], p. 196, and p. 203

Boesch, F.T.

[B10]   The strongest monotone degree condition for n-connectedness of a graph. *J. Combin. Theory* **16B** (1974) 162-165.   [71]

[B11]   On unreliability polynomials and graph connectivity in reliable network synthesis. *J. Graph Theory* **10** (1986) 339-352.   [62]

[B12]   Synthesis of reliable networks. *IEEE Trans. on Reliability* (1986) 240-246.   [62,73]

[BS5]   (with C.L. Suffel) Realizability of p-point graphs with prescribed minimum degree, maximum degree, and line-connectivity. *J. Graph Theory* **4** (1980) 363-370.   [61]

[BS6]   (with C.L. Suffel) Realizability of p-point graphs with prescribed minimum degree, maximum degree, and point-connectivity. *Discrete Applied Math.* **3** (1981) 9-18.   [61]

[BT2]   (with R. Tindell) Circulants and their connectivities. *J. Graph Theory* **8** (1984) 487-499.   [73]

[BT3]   (with R. Tindell) Connectivity and symmetry in graphs. *Graphs and Applications: Proc. First Colorado Graph Theory and Applications Symposium*, Wiley, New York (1985) 53-67.   [73–74]

[BW2]   (with J. Wang) Super line-connectivity properties of circulant graphs. *SIAM J. Algebraic and Discrete Methods* **7** (1986) 89-98.   [73]

[BW3]   (with J. Wang) Reliable circulant networks with minimum trans-
             mission delay. *IEEE Trans. Circuits Syst.* **CAS32** (1985) 1287-
             1291.   [74]

             see also [BBB1]

Bollobás, B.

[B13]   On graphs with at most three independent paths connecting any
             two vertices. *Studia Sci Math. Hungar.* **1** (1966) 137-140.   [68]

[B14]   Cycles modulo k. *Bull London Math. Soc.* **9** (1977) 97-98.   [113]

[B15]   *Extremal Graph Theory*. Academic Press, New York (1978).
                                                                         [95,105]

[B16]   The diameter of random graphs. *Trans. Amer. Math. Soc.* **267**
             (1981) 41-52.   [100,109]

[BH1]   (with F. Harary) The trail number of a graph. *Ann. Discrete Math.*
             13 (1982) 51-60.   [110]

             see also [BB1]

Bond, J.   see [BBPP1]

Bondy, J.A.

[B17]   Properties of graphs with constraints on degrees. *Studia Sci. Math.
             Hungar.* **4** (1969) 473-475.   [71]

[B18]   Pancyclic graphs. *J. Combin. Theory* **11B** (1971) 80-84.   [91]

[B19]   A remark on two sufficient conditions for hamiltonian cycles. *Dis-
             crete Math.* **22** (1978) 191-194.   [85]

[BC1]   (with V. Chvátal) A method in graph theory. *Discrete Math.* **15**
             (1976) 111-136.   [80,250]

[BK1]   (with M. Kouider) Hamiltonian cycles in regular 2-connected
             graphs. *J. Combin. Theory* **44B** (1988) 177-186.   [93]

[BT4]   (with C. Thomassen) A short proof of Meyniel's theorem. *Discrete
             Math.* **19** (1977) 195-197.   [218]

Bosák, J.

[BKZ1]   (with A. Kotzig and Š. Znám) Strongly geodetic graphs. *J. Combin.
             Theory* **5** (1968) 170-176.   [149]

[BRZ1]   (with A. Rosa and Š. Znám) On decompositions of complete graphs
             into factors with given diameters. *Theory of Graphs* (Proc. Colloq.
             Tihany), Academic Press, New York (1968) 37-56.   [106]

294

Bose, R.C.

[B20]   Strongly regular graphs, partial geometries and partially balanced
             designs. *Pacific J. Math.* **13** (1963) 389-419.   [171]

Boshier, A.G.   see [BBS1]

Bott, R.

[BM2]   (with J.P. Mayberry) Matrices and trees. *Economic Activity Anal-
             ysis*, New York (1954) 391-400.   [215]

Bourgeois, B.A.

[BHK1]   (with A.M. Hobbs and J. Kajiraj) Packing trees in complete graphs.
             Discrete Math. **67** (1987) 27-42.   [103]

Brigham, R.C.

[BD1]   (with R.D. Dutton) A compilation of relations between graph in-
             variants. *Networks* **15** (1985) 73-107.   [107]

Brooks, R.L.

[BSST1]   (with C.A.B. Smith, A.H. Stone, and W.T. Tutte) The dissection
             of rectangles into squares. *Duke Math. J.* **7** (1940) 312-340. [127]

de Bruijn, N.G.

[BE1]   (with T. van Aardenne-Ehrenfest) Circuits and trees in oriented
             graphs. *Simon Stevin* **28** (1951) 203-217.   [215]

Buckley, F.

[B21]   Self-centered graphs with a given radius. *Congressus Numer.* **23**
             (1979) 211-215.   [38]

[B22]   Mean distance in line graphs. *Congressus Numerantium* **32** (1981)
             153-162.   [193]

[B23]   The central ratio of a graph. *Discrete Math.* **38** (1982) 17-21.
                                                                                   [38,56]

[B24]   The common neighbor distribution of a graph. *Congressus Numer-
             antium* **35** (1982) 117-130.   [199,201]

[B25]   Equalities involving certain graphical distributions. *Lecture Notes
             in Mathematics* **1073** (1984) 179-192.   [193,197,201-202]

[B26]   Antigeodetic graphs. *Notes from Graph Theory Day* **7** (1984)
             19-21.   [198]

[B27]   Facility location problems. *College Mathematics Journal* **18** (1987)
             24-32.   [276]

[B28]   Self-centered graphs. *Proceedings of First China-USA International Conference on Graph Theory and its Applications*, New York Academy of Sciences, New York (to appear).   [57]

[BDH1]   (with L.L.Doty and F. Harary) On graphs with signed inverses. *Networks* **18** (1988) 151-157.   [120]

[BH2]   (with F. Harary) Closed geodetic games for graphs. *Congressus Numerantium* **47** (1985) 131-138.   [139]

[BH3]   (with F. Harary) Geodetic games for graphs. *Quaestiones Math.* **8** (1986) 321-334.   [138]

[BH4]   (with F. Harary) On the euclidean dimension of a wheel. *Graphs and Combinatorics* **4** (1988) 23-30.   [25,143,146]

[BH5]   (with F. Harary) On longest induced paths in graphs. *Chinese Quart. J. Math.* **3** (1988) 61-65.   [111]

[BHQ1]   (with F. Harary and L.V. Quintas) Extremal results on the geodetic number of a graph. *Scientia* (to appear).   [137-138]

[BL1]   (with M. Lewinter) A note on graphs with diameter-preserving spanning trees. *J. Graph Theory* **12** (1988) 525-528.   [151,235]

[BL2]   (with M. Lewinter) On graphs with center-preserving spanning trees. *Notes from Graph Theory Day* **14** (1987) 33-35.   [152]

[BL3]   (with M. Lewinter) Minimal graph embeddings, eccentric vertices, and the peripherian. *Proceedings of the Fifth Caribbean Conference on Combinatorics and Computing*, University of the West Indies (1988) 72-84.   [57]

[BL4]   (with M. Lewinter) Graphs with all diametral paths through distant central nodes. *Computers and Math. with Appl.* (to appear).
                                                                       [110]

[BMS1]   (with Z. Miller and P.J. Slater) On graphs containing a given graph as center. *J. Graph Theory* **5** (1981) 427-434.   [41]

[BP1]   (with Z. Palka) Random graphs with the distance-hereditary property. *Congressus Numerantium* **58** (1987) 217-224.   [152-154]

[BP2]   (with Z. Palka) On the center and periphery of a random graph. *Ars Combinatoria* (to appear).   [109]

[BP3]   (with Z. Palka) Preserving spanning trees in random graphs. *Proceedings of the Third International Symposium on Random Graphs and Probabilistic Methods in Combinatorics*, Wiley, New York (to appear).   [152]

[BS7]   (with L. Superville) Distance distributions and mean distance problems. *Proceedings of the Third Caribbean Conference on Combinatorics and Computing*, University of the West Indies (1981) 67-76.
                                                                    [188-190,203]
         see also p. 186

Buneman, P.

[B29]   A note on the metric properties of trees. *J. Combin. Theory* **17B**
         (1974) 48-50.   [130]

Burtin, Y.

[B30]   On extreme metric parameters of a random graph I. Asymptotic
         estimates. *Theory Prob. Appl.* **19** (1974) 710-725.   [109]

[B31]   On extreme metric characterizations of a random graph II. Limit
         Distributions. *Theory Prob. Appl.* **20** (1974) 83-101.   [109]

Caccetta, L.

[CS1]   (with W.F. Smyth) A characterization of edge-maximal diameter-
         critical graphs. (preprint)   [108]

Cameron, P.J.

[C1]   6-transitive graphs. *J. Combin.Theory* **28B** (1980) 168-179. [168]

[CV1]   (with J.H. Van Lint) *Graphs, Codes, and Designs*, Cambridge University Press, Cambridge (1980).   [171]

Camion, P.

[C2]   Chemie et circuits hamiltoniens des graphes complets. *C.R. Acad.
         Sci. Paris* **249** (1959) 2151-2152.   [222]

Capobianco, M.

[C3]   Statistical inference in finite populations having structure. *Trans.
         New York Acad. Sci., Ser. II* **32** (1970) 401-413.   [194]

[C4]   Collinearity, distance, and connectedness of graphs. *The Theory
         and Applications of Graphs*, Wiley, New York (1981) 169-179.
                                                                    [191-193]
[CM1]   (with J.C. Molluzzo) *Examples and Counterexamples in Graph
         Theory*, North Holland, Amsterdam (1978).   [63]
         see also p. 186 and p. 197

Cartwright, D.

[CH1]   (with F. Harary) The number of lines in a digraph of each con-
         nectedness category. *SIAM Review* **3** (1961) 309-314.   [209]
         see also [HNC1]

Cayley, A.

[C5]   The theory of groups, graphical representations. *Mathematical Papers*, Cambridge **10** (1895) 26-28.   [159]

Chachra, V.

[CGM1]   (with P.M. Ghare and J.M. Moore) *Applications of Graph Theory Algorithms*. North Holland, New York (1979).   [279]

Chartrand, G.

[C6]   A graph-theoretic approach to a communications problem. *J. SIAM Appl. Math* **14** (1966) 778-781.   [61]

[C7]   On hamiltonian line graphs. *Trans. Amer. Math. Soc.* **134** (1968) 559-566.   [89]

[CH2]   (with F. Harary) Planar permutation graphs *Ann. Inst. Henri Poincaré* Sec. **B3** (1967) 433-438.   [36]

[CH3]   (with F. Harary) Graphs with prescribed connectivities. *Theory of Graphs*, Akadémiai Kiadó, Budapest (1968) 61-63.   [61]

[CK1]   (with H.V. Kronk) Randomly traceable graphs. *J. SIAM Appl. Math.* **16** (1968) 696-700.   [90,93]

[CKL1]   (with S.F. Kapoor and D.R Lick) *n*-Hamiltonian graphs *J. Combin. Theory* **9** (1970) 308-312.   [85]

[CKL2]   (with A. Kaugars and D.R. Lick) Critically *n*-connected graphs. *Proc. Amer. Math. Soc.* **32** (1972) 63-68.   [20,72,75]

[CL1]   (with L. Lesniak) *Graphs & Digraphs*. Prindle, Weber, and Schmidt, Boston (1986).   [166]

[COTZ1]   (with O.R. Oellermann, S. Tian, and H.B. Zou) Steiner distance in graphs. *Casopis Pest. Mat.* (to appear).   [144-145]

[CN1]   (with E.A. Nordhaus) Graphs hamiltonian-connected from a vertex. *The Theory and Applications of Graphs*, Wiley, New York (1981) 189-201.   [94]

[CS2]   (with M.J. Stewart) The connectivity of line-graphs. *Math. Ann.* **182** (1969) 170-174.   [62,90]

[CW1]   (with C.E. Wall) On the hamiltonian index of a graph. *Studia Sci. Math. Hungar.* **8** (1973) 43-48.   [88]

Chen, Z-b

[C8]   On the existence of graphs with A(H) = 3. [in Chinese] *J. Math. (PRC)* **4** (3) (1984) 267-271.   [41]

Cheng, Y.

[CH4]   (with F.K. Hwang) Diameters of weighted double loop networks.
J. Algorithms **9** (1988) 401-410.   [218]

see also p. 218

Chinn, P.Z.

[C9]   The path centrix of a tree. *Proceedings of the Ninth Southeast-
ern Conference on Combinatorics, Graph Theory, and Computing,*
Utilitas Mathematica, Winnipeg (1978) 195-202.   [56]

[CM2]   (with B.C. Marcot) Using graph theory measures for assessing di-
versity of wildlife habitat. *Research Report 83-1*, Humboldt State
University (1983) 44 pages.   [7]

Christofides, N.

[C10]   *Graph Theory: An Algorithmic Approach.* Academic Press, New
York (1975).   [276]

Chung, F.R.K

[C11]   The average distance and the independence number. *J. Graph Theory* **12** (1988) 229-235.   [204]

[C12]   Diameters and eigenvalues. *J. Amer. Math. Soc.* **2** (1989) 187-196.
    [134]

Chvátal, V.

[C13]   On hamilton's ideals. *J. Combin. Theory* **12B** (1972) 163-168.
    [79]

[C14]   Flip-flops in hypohamiltonian graphs. *Canad. Math. Bull.* **16**
(1973) 33-41.   [94]

[C15]   Tough graphs and hamiltonian circuits. *Discrete Math.* **5** (1973)
215-228.   [112]

[CE1]   (with P. Erdős) A note on hamiltonian circuits. *Discrete Math.* **2**
(1972) 111-113.   [84]

see also [BC1]

Clark, L.H.   see [BCDEF1]

Cockayne, E.J.

[CHH1]   (with S.M. Hedetmiemi and S.T. Hedetniemi) Linear algorithms
for finding the Jordan center and path center of a tree. *Transportation Sci.* **15** (1981) 98-114.   [47,242]

Colbourn, C.J.

[C16]   *The Combinatorics of Network Reliability.* Oxford University
Press, New York (1987).   [76]

Cook, R.J.

[CP1]   (with D.G. Price) A class of geodetic blocks. *J. Graph Theory* **6** (1982) 266-280.   [149]

Conradt, D.

[CP2]   (with U. Pape) Maximales Matching in Graphen. *Ausgewählte Operations Research Software in Fortran*, Oldenburg, Munich (1980) 103-114.   [244,247]

Dauber, E.   see p. 163

Dean, N.

[D1]   How do you decompose a graph into trees of small diameter? *Congressus Numer.* **62** (1988) 65-67.   [105]

Delire, J.M.

[D2]   Graphs with high radon number. *Acad. Royal Belg. Bull. Cl. Sci.* **70** (1984) 14-24.   [140]

Deo, N.   see [RND1] and [SDK1]

Dijkstra, E.W.

[D3]   A note on two problems in connection with graphs. *Numeriske Math* **1** (1959) 269-271.   [270]

Dilworth, R.P.

[D4]   A decomposition theorem for partially ordered sets. *Ann. Math.* **51** (1950) 161-166.   [75]

Dirac, G.A.

[D5]   Some theorems on abstract graphs. *Proc. London Math. Soc., Ser. 3* **2** (1952) 69-81.   [79]

[D6]   In abstrakten Graphen verhandene vollständige 4-Graphen und ihre Unterteilungen. *Math. Nachr.* **22** (1960) 61-85.   [62]

[D7]   Généralizations du théorème de Menger. *C.R. Acad. Sci. Paris* **250** (1960) 4252-4253.   [68]

[D8]   Short proof of Menger's graph theorem. *Mathematika* **13** (1966) 42-44.   [64]

[D9]   Minimally 2-connected graphs. *J. Reine Angew. Math.* **228** (1967) 204-216.   [104]

Djokovič, D.Z.

[D10]   Distance preserving subgraphs of hypercubes. *J. Combin. Theory* **14B** (1973) 263-267.   [143]

Douthett, J.   see [BCDEF1]

Doty, L.L.   see [BDH1]

Doyle, J.K.

[DG1]   (with J.E. Graver) Mean distance in a graph. *Discrete Math.* **17**
             (1977) 147-154.   [189,191]

[DG2]   (with J.E. Graver) Mean distance in a directed graph. *Environ-
             ment and Planning* **5B** (1978) 19-29.   [193]

Dutton, R.D.   see [BD1]

Ebert, J.

[E1]   A linear disjoint path algorithm. *Proceedings of the 8th Conference
             on Graphtheoretic Concepts in Computer Science.* Hanser Verlag,
             Munich (1982) 37-45.   [235]

Edelberg, M.

[EGG1]   (with M.R. Garey and R.L. Graham) On the distance matrix of a
             tree. *Discrete Math.* **14** (1976) 23-39.   [131-132]

Edmonds, J.

[E2]   Paths, trees, and flowers. *Canad. J. Math.* **17** (1965) 449-467.
                                                                            [244]

[EK1]   (with R.M. Karp) Theoretic improvements in algorithmic efficiency
             for network flow problems. *J. Assoc. Comput. Mach.* **19** (1972) 248-
             264.   [257]

van Aardenne-Ehrenfest, T.   see [BE1]

Elias, P.

[EFS1]   (with A. Feinstein and C.E. Shannon) A note on the maximum
             flow through a network. *IRE Trans. Inform. Theory* **IT-2** (1956)
             117-119.   [66]

Entringer, R.C.

[EJS1]   (with D.E. Jackson and D.E. Snyder) Distance in graphs. *Czech.
             Math. J.* **26** (1976) 283-296.   [43-44,46,184]

[EJS2]   (with D.E. Jackson and P.J. Slater) Geodetic connectivity of
             graphs. *IEEE Trans. Circuits Syst.* **CAS24** (1977) 460-463. [107]

[ES1]   (with P.J. Slater) A theorem on critically 3-connected graphs.
             Nanta Math. **11** (1978) 141-145.   [75]

             see also [BCDEF1] and p. 184

Erdős, P.

[EFGS1]   (with R.J. Faudree, A. Gyárfás, and R.H. Schelp) Odd cycles in graphs of given minimum degree. *Graph Theory, Combinatorics, and Applications*, Wiley, New York (to appear)   [114]

[EFH1]   (with S. Fajtlowicz and A.J. Hoffman) Maximum degree in graphs of diameter 2. *Networks* **10** (1980) 87-90.   [104]

[EG1]   (with T. Gallai) Graphs with prescribed degrees of vertices. [in Hungarian] *Mat. Lapok* **11** (1960) 264-274.   [173]

[EHT1]   (with F. Harary and W.T. Tutte) On the dimension of a graph. *Mathematika* **12** (1965) 118-122.   [143]

[ES2]   (with H. Sachs) Reguläre Graphen gegebener Taillenweite mit minimaler Knotenzahl. *Wiss. Z. Univ. Halle, Math.-Nat.* **12** (1963) 251-258.   [165]

            see also [CE1]

Euler, L.

[E3]   Solutio problematis ad geometriam situs pertinentis. *Comment. Academiae Sci. I. Petropolitanae* **8** (1736) 128- 140. *Opera Omnia Series I-7* (1766) 1-10.   [3]

Even, S.

[E4]   *Graph Algorithms.* Computer Science Press, Potomac, MD (1979).   [230]

Everett, M.G.

[ES3]   (with S.B. Seidman) The hull number of a graph. *Disc. Math.* **57** (1985) 217-223.   [139]

Exoo, G.

[EHX1]   (with F. Harary and C. Xu) On vulnerability in graphs of diameter four. *Comput. Math. Appl.* (to appear).   [76]

Fajtlowicz, S.   see [EFH1]

Fan, G.H.

[F1]   New sufficient conditions for cycles in graphs. *J. Combin. Theory* **37B** (1984) 221-227.   [82,113]

Farber, M.

[FJ1]   (with R.E Jamison) Convexity in graphs and hypergraphs. *SIAM J. Alg. Disc. Meth.* **7** (1986) 433-444.   [140,154]

Farrell, E.J.

[F2]   On a class of polynomials obtained from the circuits in a graph and its application to characteristic polynomials of graphs. *Discrete Math.* **25** (1979) 121-133.   [134]

Faudree, R.J.

[FGJS1]   (with R.J. Gould, M.S. Jacobson, and R.H. Schelp) Neighborhood unions and hamiltonian properties of graphs. *J. Combin. Theory* **47B** (1989) 1-9.   [80-81]

[FRS1]   (with C.C. Rousseau and R.H. Schelp) Theory of path length distributions I. *Discrete Math.* **6** (1973) 35-52.   [195]

[FS1]   (with R.H. Schelp) Path connected graphs. *Acta Math. Acad. Sci. Hungar.* **25** (1974) 313-319.   [195]

[FS2]   (with R.H. Schelp) Various length paths in graphs. *Theory and Applications of Graphs*, Springer-Verlag, Berlin (1978) 160-173.   [195]

see also [EFGS1]

Feinstein, A.   see [EFS1]

Fellows, M.R.

[FFL1]   (with D.K. Friesen and M.A. Langston) On finding optimal and near-optimal lineal spanning trees. *Algorithmica* **3** (1988) 549-560.   [235]

see also [BCDEF1]

Fink, J.F.

[F3]   A note on unilateral digraphs. *Bull. Calcutta Math. Soc.* **73** (1981) 9-10.   [224]

Fleischner, H.

[F4]   The square of every two-connected graph is hamiltonian. *J. Combin. Theory* **16B** (1974) 29-34.   [86]

Floyd, R.W.

[F5]   Algorithm 97: Shortest paths. *Comm. ACM* **5** (1962) 345.   [271]

Ford, L.R.

[FF1]   (with D.R. Fulkerson) Maximal flow through a network. *Canad. J. Math.* **8** (1956) 399-404.   [66,256-258]

[FF2]   (with D.R. Fulkerson) *Flows in Networks*, Princeton University Press, Princeton (1962).   [256]

Foster, R.M.

[F6]   Geometric circuits of electrical networks. *Trans. Amer. Inst. Elec.*
            Engrs. **51** (1932) 309-317.   [162]

Fouldes, L.R.

[F7]   Enumeration of graph theoretic solutions for facilities layout. *Con-*
            gressus Numer. **48** (1985) 87-99.   [5]

[FR1]   (with D.R. Robinson) *Digraphs: Theory and Techniques.* Gordon
            & Breach, New York (1980).   [205]

Foulkes, J.D.

[F8]   Directed graphs and assembly schedules. *Proc. Symp. Appl. Math.*
            10, Amer. Math. Soc. (1960) 281-289.   [222]

Fraisse, P.

[F9]   A new sufficient condition for hamiltonian graphs. *J. Graph Theory*
            10 (1986) 405-409.   [81]

Freeman, L.C.

[F10]   A set of measures of centrality based on betweenness. *Sociometry*
            40 (1977) 35-41.   [55]

Friesen, D.K.   see [FFL1]

Frucht, R.

[F11]   Herstellung von Graphen mit vorgegebenen abstracten Gruppe.
            Compositio Math. **6** (1938) 230-250.   [159]

[F12]   Graphs of degree three with a given abstract group, *Canad. J.*
            Math. **1** (1949) 365-378.   [161]

[F13]   On the groups of repeated graphs. *Bull. Amer. Math. Soc.* **55**
            (1949) 418-420.   [161]

[FH1]   (with F. Harary) On the corona of two graphs. *Aequationes Math.*
            4 (1970) 322-324.   [24,161]

Fulkerson, D.R.

[F14]   Networks, frames, and blocking systems. *Mathematics of the De-*
            cision Sciences, Part 1, (1968) 303-334.   [69]

            see also [FF1] and [FF2]

Gabow, H.

[G1]   An efficient implementation of Edmond's algorithm for maximum
            matchings on graphs. *J. ACM* **23** (1975) 221-234.   [244]

Gallai, T.

[G2]   On directed paths and circuits. *Theory of Graphs*, Academic Press, New York (1968) 115-119.   [217]

see also [EG1]

Garey, M.R.

[GJ1]   (with D.S. Johnson) *Computers and Intractability: A Guide to the Theory of NP-Completeness*, Freeman, San Francisco (1979).   [229-230,249]

see also [EGG1]

Gaudin, T.

[GHR1]   (with J.C. Herz and P. Rossi) Solution du problème no. 29. *Rev. Franc. Rech. Oper.* **8** (1964) 214-218.   [94]

Germa, A.   see [BGHS1]

Gewirtz, A.

[G3]   Graphs of maximal even girth. *Canadian J. Math.* **21** (1969) 915-934.   [171]

Ghare, P.M.   see [CGM1]

Ghoula-Houri, A.

[G4]   Une condition suffisante d'éxistence d'un circuit hamiltonien. *C.R. Acad. Sci. Paris* **251** (1960) 495-497.   [219]

Gibbons, A.

[G5]   *Algorithmic Graph Theory*. Cambridge University Press, London (1985).   [265]

Gliviak, F.

[G6]   On certain classes of graphs of diameter two without superfluous edges. *Acta Fac. R.N. Univ. Comen., Math.* **21** (1968) 39-48.   [101,104]

[G7]   On radially critical graphs. in *Recent Advances in Graph Theory*, *Proc. Sympos. Prague*, Academia Praha, Prague (1975) 207-221.   [97-98,100]

[G8]   On the structure of radially critical graphs. in *Graphs, Hypergraphs and Block Systems*, Zielona Góra, Poland (1976) 69-73.   [98]

[G9]   Vertex-critical graphs of given diameter. *Acta Math. Acad. Sci. Hungar.* **27** (1976) 255-262.   [98,102,105-106]

Godsil, C.D.

[GS1]   (with J. Shawe-Taylor) Distance-regularised graphs are distance-regular or distance-biregular. *J. Combin. Theory* **43B** (1987) 14-24.   [170]

Goodman, S.E.

[GH1]   (with S.T. Hedetniemi) Eulerian walks in graphs. *SIAM J. Comput.* **2** (1973) 16-27.   [273]

Gould, R.J.

[G10]   *Graph Theory*. Benjamin Cummings, Menlo Park (1988).
                                                                   [247,258,265]

[GL1]   (with T.E. Lindquester) Some extremal problems involving adjacency conditions for vertices at distance two. *Recent Studies in Graph Theory*, Vishwa International, Gulbarga, India (1989) 140-148.   [81]

          see also [FGJS1]

Graham, N.

[GH2]   (with F. Harary) Changing and unchanging the diameter of a hypercube. *Discrete Appl. Math.* (submitted)   [115]

Graham, R.L.

[GHH1]   (with A.J. Hoffman and H. Hosoya) On the distance matrix of a directed graph. *J. Graph Theory* **1** (1977) 85-88.   [133]

[GL2]   (with L. Lovász) Distance matrix polynomials of trees. *Adv. Math.* **29** (1978) 60-88.   [132]

[GP1]   (with H.O. Pollack) On the addressing problem for loop switching. *Bell System Tech. J.* **50** (1971) 2495-2519.   [132]

          see also [EGG1]

Graver, J.E.   see [DG1] and [DG2]

Gu, W.

[GR1]   (with K.B. Reid) Peripheral and eccentric vertices in graphs. (preprint).   [57]

Guan, M.G.

[G11]   Graphic programming using odd or even points. *Chinese Math.* **1** (1962) 273-277.   [272]

Gyárfás, A.   see [EFGS1]

Hage, P.

[HH1]   (with F. Harary) *Structural Models in Anthropology.* Cambridge University Press, Cambridge (1983).   [55]

[HH2]   (with F. Harary) *Exchange in Oceania*, Oxford University Press, London (submitted).   [55]

Hakimi, S.L.

[H1]   On the realizability of a set of integers as the degrees of the vertices of a graph. *J. SIAM Appl. Math.* **10** (1962) 496-506.   [173]

[H2]   Optimal locations of switching centers and the absolute centers and medians of graphs. *Operations Research* **12** (1964) 450-459.
                                                                         [276]

[HY1]   (with S.S. Yau) Distance matrix of a graph and its realizability. *Quart. Appl. Math* **22** (1964) 305-317.   [129]

Halberstam, F.   see p. 191

Hall, P.

[H3]   On representations of subsets. *J. London Math. Soc.* **10** (1935) 26-30.   [70]

Handler, G.Y.

[HM1]   (with P.B. Mirchandani) *Location on Networks: Theory and Algorithms.* The MIT Press, Cambridge, Mass. (1979).   [242,276]

Harary, F.

[H4]   On the notion of balance of a signed graph. *Mich. Math. J.* **2** (1953) 143-146.   [120-122]

[H5]   Structural duality. *Behavioral Sci.* **2** (1957) 255-265.   [30]

[H6]   Status and contrastatus. *Sociometry* **22** (1959) 23-43.   [42,224]

[H7]   An elementary theorem on graphs. *Amer. Math. Monthly* **66** (1959) 405-407.   [21]

[H8]   On the group of the composition of two graphs. *Duke Math. J.* **26** (1959) 29-34.   [23]

[H9]   Who eats whom? *General Systems* **6** (1961) 41-44.   [7]

[H10]   A structural analysis of the situation in the Middle East in 1956. *J. Conflict Resolution* **5** (1961) 167-178.   [120]

[H11]   The determinant of the adjacency matrix of a graph. *SIAM Review* **4** (1962) 202-210.   [120]

[H12]   The maximum connectivity of a graph. *Proc. Nat. Acad. Sci. U.S.A.*
             48 (1962) 1142-1146.   [61]

[H13]   On the reconstruction of a graph from a collection of subgraphs.
             Theory of Graphs and its Applications, Academic Press, New York
             (1964) 47-52.   [11]

[H14]   *Graph Theory*. Addison-Wesley, Reading, Mass. (1969).
                                                               [2,23,30,213]

[H15]   Conditional connectivity. *Networks* **13** (1983) 347-357.   [74]

[H16]   Convexity in graphs: achievement and avoidance games. *Ann. Dis-
             crete Math.* **20** (1983) 323.   [138-139]

[HKMR1]   (with C. King, A. Mowshowitz, and R.C. Read) Cospectral graphs
             and digraphs. *Bull. London Math. Soc.* **3** (1971) 321-328.   [122]

[HK1]   (with P. Kovaks) Regular graphs with given girth pair. *J. Graph
             Theory* **7** (1983) 209-218.   [166]

[HL1]   (with D.E. Leep) Angles in graphs. *Two-Year College Math. Read-
             ings*, Math. Assoc. Amer., Washington (1981) 62-74.   [192]

[HM2]   (with B. Manvel) On the number of cycles in a graph. *Mat. Časopis
             Sloven Akad. Vied* **21** (1971) 55-63.   [133]

[HM3]   (with R.A. Melter) On the metric dimension of a graph. *Ars Com-
             binatoria* **2** (1976) 191-195.   [144,146]

[HM4]   (with R.A. Melter) The graphs with no equilateral triangles. *Gazeta
             Math.* **3** (1982) 182-183.   [146,192]

[HMPT1]   (with R.A. Melter, U.N. Peled, and I. Tomescu) Boolean distance
             for graphs. *Discrete Math.* **39** (1982) 123-127.   [144]

[HM5]   (with H. Minc) Which nonnegative matrices are self-inverse? *Math.
             Mag.* **49** (1976) 91-92.   [120]

[HM6]   (with L. Moser) The theory of round robin tournaments. *Amer.
             Math. Monthly* **73** (1966) 231-246.   [223-224]

[HN1]   (with C.St.J.A. Nash-Williams) On eulerian and hamiltonian
             graphs and line-graphs. *Canad. Math. Bull.* **8** (1965) 701-709. [89]

[HN2]   (with J. Nieminen) Convexity in graphs. *J. Diff. Geometry* **16**
             (1981) 185-190.   [136]

[HN3]   (with R.Z. Norman) The dissimilarity characteristic of Husimi
             trees. *Ann. Math.* **58** (1953) 134-141.   [34]

[HN4]   (with R.Z. Norman) Some properties of line digraphs. *Rend. Circ.
             Mat. Palermo* **9** (1961) 161-168.   [210]

[HNC1]   (with R.Z. Norman and D. Cartwright) *Structural Models: An Introduction to the Theory of Directed Graphs*. Wiley, New York (1965).   [205,237]

[HO1]   (with P. Ostrand) The cutting center theorem for trees. *Discrete Math.* **1** (1971) 7-18.   [51]

[HP1]   (with E.M. Palmer) The smallest graph whose group is cyclic. *Czech. Math. J.* **16** (1966) 70-71.   [161]

[HP2]   (with E.M. Palmer) On similar points of a graph. *J. Math Mech.* **15** (1966) 623-630.   [162]

[HP3]   (with E.M. Palmer) *Graphical Enumeration*. Academic Press, New York (1973).   [216]

[HP4]   (with M. Plantholt) *Trees* (in preparation).   [12]

[HR1]   (with R.W. Robinson) The diameter of a graph and its complement. *Amer. Math. Monthly* **92** (1985) 211-212.   [22]

[HR2]   (with I.C. Ross) The square of a tree. *Bell Syst. Tech. J.* **39** (1960) 641-647.   [26]

[HR3]   (with I.C. Ross) A description of strengthening and weakening group members. *Sociometry* **22** (1959) 139-147.   [209]

[HT1]   (with C.A. Trauth, Jr.) Connectedness of products of two directed graphs. *SIAM J. Appl. Math.* **14** (1966) 250-254.   [209]

[HT2]   (with R. Tindell) The minimal blocks of diameter two and three. *Graphs and Applications, Proceedings of the First Colorado Symposium on Graph Theory*, Wiley, New York (1985) 163-181. [104]

[HW1]   (with G. Wilcox) Boolean operations on graphs. *Math Scand.* **20** (1967) 41-51.   [30]

           see also [AH1], [BH1], [BDH1], [BH2], [BH3], [BH4], [BH5], [BHQ1], [CH1], [CH2], [CH3], [EHT1], [EHX1], [FH1], [GH2], [HH1], and [HH2]

Havel, V.

[H17]   A remark on the existence of finite graphs. [in Czech] *Časopis Pěst. Mat.* **80** (1955) 477-480.   [173]

Hebbare, S.R.

[H18]   Some properties of planar distance convex simple graphs. (preprint)   [151]

[HR4]   (with S.B. Rao) Characterization of planar distance convex simple graphs. *Proc. Symp. in Graph Theory ISI*, Calcutta (1976) 138-150.   [150]

Hedetniemi, S.M.

[HH3]   (with S.T. Hedetniemi) Centers of recursive graphs. Technical Re-
            port CS-TR-79-11, Department of Computer Science, University
            of Oregon (1979) 13 pages.   [36,38]

[HHS1]   (with S.T. Hedetniemi and P.J. Slater) Centers and medians of
            C_n-trees. Utilitas Math. 21 (1982) 225-234.   [36]

            see also [CHH1]

Hedetniemi, S.T.   see [CHH1], [GH1], [HH3], and [HHS1]

Hedman, B.

[H19]   Diameters of iterated clique graphs. Hadronic J. 9 (1986) 273-276.
                                                                   [109,115]

Hendry, G.R.T.

[H20]   Existence of graphs with prescribed mean distance. J. Graph The-
            ory 10 (1986) 173-175.   [191]

[H21]   On mean distance in certain classes of graphs. Networks 19 (1989)
            451-457.   [191,193]

[H22]   The size of graphs uniquely hamiltonian-connected from a vertex.
            Discrete Math. 61 (1986) 57-60.   [94]

Herz, J.C.   see [GHR1]

Heydemann, M.C.   see [BGHS1]

Hierholzer, C.

[H23]   Ueber die Möglichkeit, einen Linienzug Ohne Wiederholung und
            Ohne Unterbrechnung zu Umfahren. Math. Ann. 6 (1873) 30-42.
                                                                       [274]

Hobbs, A.M.   see [BHK1]

Hoffman, A.J.

[H24]   On the polynomial of a graph. Amer. Math. Monthly 70 (1963)
            30-36.   [224]

[HS1]   (with R.R. Singleton) On Moore graphs with diameters 2 and 3.
            IBM J. Res. Devel. 4 (1960) 497-504.   [101]

            see also [EFH1] and [GHH1]

Holbert, K.S.

[H25]   A note on graphs with distant center and median. Recent Studies
            in Graph Theory, Vishwa International Pub. (1989) 155-158. [57]

Hosoya, H.   see [GHH1]

Howorka, E.

[H26]   A characterization of distance-hereditary graphs. *Quart. J. Math. Oxford* **28** (1977) 417-420.   [152]

[H27]   A characterization of ptolemaic graphs. *J. Graph Theory* **5** (1981) 323-331.   [154]

Hubaut, X.

[H28]   Strongly regular graphs. *Discrete Math.* **13** (1975) 357-381. [171]

Hwang, F.K.

[HX1]   (with Y.H. Xu) Double loop networks with minimum delay. *Discrete Math.* **66** (1987) 109-118.   [218]

          see also [CH4]

Imrich, W.

[I1]   On metric properties of tree-like spaces. *Contributions to Graph Theory and its Applications*, Technische Hochschule Ilmenau, Ilmenau (1977) 129-156.   [130]

Ishizaki, Y.

[IOW1]   (with T. Ohtsuki and H. Watanabe) Topological degrees of freedom and mixed analysis of electrical networks. *IEEE Trans. Circuit Theory* **CT17** (1970) 491-499.   [236]

Jackson, B.

[J1]   Hamiltonian cycles in regular 2-connected graphs. *J. Combin. Theory* **29B** (1980) 27-46.   [93]

Jackson, D.E.   see [EJS1] and [EJS2]

Jacobson, M.S.   see [FGJS1]

Jamison, R.E.   see [FJ1]

Johnson, D.S.   see [GJ1]

Jordan, C.

[J2]   Sur les assemblages de lignes. *J. Reine Agnew. Math.* **70** (1869) 185-190.   [32,34]

Kapoor, S.F.

[KKL1]   (with H.V. Kronk and D.R. Lick) On detours in graphs. *Canad. Math. Bull.* **11** (1968) 195-201.   [111]

          see also [CKL1]

Karp, R.M.   see [EK1]

Kasiraj, J.   see [BHK1]

Kastelyn, P.W.

[K1]   Graph theory and crystal physics. *Graph Theory and Theoretical Physics*, Academic Press, London (1967) 44-110.   [216]

Kaugers, A.   see [CKL2]

Kay, D.C.

[KW1]   (with E.G. Womble) Axiomatic convexity theory and the relationship between the Carathéodory, Helly and Radon numbers. *Pacific J. Math.* **38** (1971) 471-485.   [141]

Kennedy, J.W.

[KQ1]   (with L.V. Quintas) Extremal f-trees and embedding spaces for molecular graphs. *Discrete Applied Math.* **5** (1983) 191-209. [204]

            see also [BKQ1], [BKQ2], and p. 203

King, C.   see [HKMR1]

Kirchhoff, G.

[K2]   Über die Ausflösung der Gleichungen, auf welche man bei der Untersuchung der linearen Verteilung galvanischer Ströme geführt wird. *Ann. Phys. Chem.* **72** (1847) 497-508.   [123]

Klee, V.

[KL1]   (with D. Larman) Diameters of random graphs. *Canad. J. Math.* **33** (1981) 618-640.   [100]

[KQ2]   (with H. Quaif) Classification and enumeration of minimum $(d, 1, 3)$-graphs and $(d, 2, 3)$-graphs. *J. Combin. Theory* **23B** (1977) 83-93.   [107]

Knickerbocker, C.J.

[KLS1]   (with P.F. Lock and M. Sheard) The minimum size of graphs hamiltonian-connected from a vertex. *Discrete Math* **76** (1989) 277-278.   [94]

[KLS2]   (with P.F. Lock and M. Sheard) On the structure of graphs uniquely hamiltonian-connected from a vertex. (preprint).   [94]

König, D.

[K3]   Graphen und Matrizen. *Mat. Fiz. Lapok* **38** (1931) 116-119. [70]

[K4]   *Theorie der endlichen und unendlichen Graphen.* Leipzig (1936).
        Reprinted by Chelsea, New York (1950).   [14,159]

Kotzig, A.   see [BKZ1]

Kouider, M.   see [BK1]

Kovaks, P.   see [HK1]

Kowalik, J.S.   see [SDK1]

Krausz, J.

[K5]   Démonstration nouvelle d'un théorème de Whitney sur les réseaux.
        Mat. Fiz. Lapok **50** (1943) 75-89.   [27]

Kronk, H.V.

[K6]   Generalization of a theorem of Pósa. *Proc. Amer. Math. Soc.* **21**
        (1969) 77-78.   [82]

        see also [CK1] and [KKL1]

Kruskal, J.B., Jr.

[K7]   On the shortest spanning subtree of a graph and the traveling
        salesman problem. *Proc. Amer. Math. Soc.* **7** (1956) 48-50. [262]

Kuratowski, K.

[K8]   Sur le problème des courbes gauches en topologie. *Fund. Math.* **15**
        (1930) 271-283.   [35]

Landau, H.G.

[L1]   On dominance relations and the structure of animal societies, III;
        the condition for a score structure. *Bull. Math. Biophys.* **15** (1953)
        143-148.   [223]

Langston, M.A.   see [FFL1]

Larman, D.   see [KL1]

Laskar, R.

[LS1]   (with D. Shier) On powers and centers of chordal graphs. *Discrete
         Applied Math.* **6** (1983) 139-147.   [36,38,40]

Lee, S.M.

[L2]   Design of e-invariant networks. *Congressus Numer.* **65** (1988)
        89-102.   [116]

[LW1]   (with A.Y. Wang) On critical and co-critical diameter edge-invar-
         iant networks. (preprint).   [116]

Leep, D.E.   see [HL1]

Lesniak, L.

[L3]   Eccentric sequences in graphs. *Period. Math. Hungar.* **6** (1975)
            287-293.   [176-177]

[L4]   Neighborhood unions and graphical properties. *Graph Theory,
            Combinatorics, and Applications*, Wiley, New York (to appear).
                                                                    [82,92]

            see also [CL1]

Lewinter, M.   see [BL1], [BL2], [BL3], and [BL4]

Lick, D.R.   see [CKL1], [CKL2], and [KKL1]

Lindquester, T.E.

[L5]   The effects of distance and neighborhood union conditions on
            hamiltonian properties in graphs. *J. Graph Theory* **13** (1989) 335-
            352.   [81]

            see also [GL1]

Liu, X.

[L6]   Two infinite classes of graphs with A(H) = 3. (preprint).   [41]

Liu, Z.H.

[LYZ1]   (with Z.G. Yu and Y.J. Zhu) An improvement of Jackson's re-
            sult on hamiltonian cycles in 2-connected regular graphs. *Cycles
            in Graphs* Burnaby, B.C., 1982, North-Holland, Amsterdam (1985)
            237-247.   [93]

Lock, P.F.   see [KLS1] and [KLS2]

Lovász, L.

[L7]   On coverings of graphs. *Theory of Graphs.* Academic Press, New
            York (1968) 231-236.   [103]

            see also [GL2]

Mader, W.

[M1]   On *k*-critical, *n*-connected graphs. *Progress in Graph Theory*, Aca-
            demic Press, New York (1984) 389-398.   [76]

Maehara, H.

[M2]   On the euclidean dimension of a complete multipartite graph. *Dis-
            crete Math.* **72** (1988) 285-289.   [143]

Manvel, B.   see [HM2]

March, L.

[MS1]   (with P. Steadman) *The Geometry of the Environment.* Royal Institute of British Architects, London (1971), Ch. 14.   [189]

Marcot, B.C.   see [CM2]

Maurer, S.B.

[MS2]   (with P.J. Slater) On k-critical n-connected graphs. *Discrete Math.* **20** (1977) 255-262.   [75]

Mayberry, J.P.   see [BM2]

McCracken, D.D.

[M3]   *A Second Course in Computer Science with Pascal.* Wiley, New York (1987).   [6]

McKay, B.D.

[M4]   On the spectral characterisation of trees. *Ars Combinatoria* **3** (1977) 219-232.   [132]

Melter, R.A.

[MT1]   (with I. Tomescu) Isometric embeddability for graphs. *Ars Combinatoria* **12** (1981) 111-115.   [142]

[MT2]   (with I. Tomescu) On the Boolean metric dimension of a graph. *Rev. Roumaine Math. Pures Appl.* **29** (1984) 407-415.   [144]

see also [HM3], [HM4], and [HMPT1]

Menger, K.

[M5]   Zur allgemeinen Kurventheorie. *Fund. Math.* **10** (1927) 96-115.   [64]

Meyniel, M.

[M6]   Une condition suffisante d'éxistence d'un circuit hamiltonien dans un graph oriente. *J. Combin. Theory* **14B** (1973) 137-147.   [218]

Micali, S.

[MV1]   (with V.V. Vazirani) An $O(\sqrt{|V|} \cdot |E|)$ algorithm for finding maximum matchings in general graphs. *Proc. 21st Annual Symp. on Foundations of Computer Science*, IEEE, Long Beach (1980) 17-27.   [247]

Miller, Z.

[M7]   Medians and distance sequences in graphs. *Ars Combin.* **15** (1983) 169-177.   [45]

see also [BMS1]

Minc, H.   see [HM5]

Minieka, E.

[M8]   *Optimization Algorithms for Networks and Graphs.* Wiley, New
            York (1978).   [242]

Mirchandani, P.B.   see [HM1]

Molluzzo, J.C.   see [CM1]

Moon, J.W.

[M9]   On subtournaments of a tournament. *Canad. Math. Bull.* **9** (1966)
            297-301.   [223]

[M10]   Various proofs for Cayley's formula for counting trees. *A Seminar
            in Graph Theory,* Holt, Rinehart and Winston, New York (1967)
            70-78.   [125]

[M11]   *Topics on Tournaments.* Holt, Rinehart, and Winston, New York
            (1968).   [205]

[M12]   On a problem of Ore. *Math. Gazette* **49** (1965) 40-41.   [93]

[MM1]   (with L. Moser) Almost all $(0,1)$-matrices are primitive. *Studia Sci.
            Math. Hungar.* **1** (1966) 153-156.   [100]

Moore, J.M.   see [CGM1]

Morgan, C.A.

[MS3]   (with P.J. Slater) A linear algorithm for the core of a tree. *J. Al-
            gorithms* **1** (1980) 247-258.   [242]

Morgana, A.   see [BMSV1]

Moser, L.   see [HM6] and [MM1]

Mowshowitz, A.

[M13]   The characteristic polynomial of a graph. *J. Combin. Theory* **12B**
            (1972) 177-193.   [131]

            see also [HKMR1]

Mulder, H.M.   see [BM1]

Myers, B.R.

[M14]   The Klee and Quaife minimum $(d,1,3)$-graphs revisited. *IEEE
            Trans. Circuits Syst.* **27** (1980) 214-220.   [107]

[M15]   The minimum-order three-connected cubic graphs with specified
            diameters. *IEEE Trans. Circuits Syst.* **27** (1980) 698-709. [107]

Nandakumar, R.

[N1]   *On Some Eccentricity Properties of Graphs.* Ph.D. Thesis, Indian
         Institute of Technology, India (1986).   [99-100,151,177,179]

[NP1]   (with K.R. Parthasarathy) Unique eccentric point graphs. *Discrete
         Math.* **46** (1983) 69-74.   [37-38,40,42]

Nash-Williams, C.St.J.A.

[N2]   Hamiltonian arcs and circuits. *Recent Trends in Graph Theory,*
         Springer-Verlag, Berlin (1971) 197-210.   [85]

         see also [HN1]

Necásková, M.

[N3]   A note on the achievement geodetic games. *Quaestiones Math.* **12**
         (1988) 115-119.   [138]

Nieminen, J.

[N4]   Characterizations of graphs by convex hulls. *Bull. Inst. Math.
         Acad. Sinica* **10** (1982) 177-184.   [136]

         see also [HN2]

Nievergelt, J.   see [RND1]

Nomura, K.

[N5]   An inequality between intersection numbers of a distance-regular
         graph. *J. Combin. Theory* **43B** (1987) 358-359.   [170]

Nordhaus, E.A.   see [CN1]

Norman, R.Z.   see [HN3], [HN4], and [HNC1]

Oellermann, O.R.

[OT1]   (with S. Tian) Steiner centers in graphs. *J. Graph Theory* (to
         appear).   [145]

         see also [COTZ1]

Ohtsuki, T.   see [IOW1]

Opatrny, J.   see [AOS1]

Ore, O.

[O1]   Note on hamiltonian circuits. *American Math. Monthly* **67** (1960)
         55.   [79]

[O2]   Arc coverings of graphs. *Ann. Mat. Pura Appl.* **55** (1961) 315-322.
                                                        [82]

[O3]   Diameters in graphs. *J. Combin. Theory* **5** (1968) 75-81.
                                                                    [36,106,108]

[O4]   Hamilton connected graphs. *J. Math. Pures Appl.* **42** (1963)
            21-27.   [92]

Ostrand, P.A.

[O5]   Graphs with specified radius and diameter. *Discrete Math.* **4** (1973)
            71-75.   [176]

            see also [HO1]

Overbeck-Larisch, M.

[O6]   Hamiltonian paths in oriented graphs. *J. Combin. Theory* **21B**
            (1976) 76-80.   [220-221]

Palka, Z.   see [BP1], [BP2], [BP3]

Palmer, E.M.

[P1]   *Graphical Evolution: An Introduction to the Theory of Random
            Graphs.* Wiley, New York (1985).   [105]

            see also [IIP1], [IIP2], [HP3]

Paoli, M.   see [BBPP1]

Pape, U.   see [CP2]

Parthasarathy, K.R.

[PS1]   (with N. Srinivasan) Some general constructions of geodetic blocks.
            J. Combin. Theory **33B** (1982) 121-136.   [147,149-150]

[PS2]   (with N. Srinivasan) Geodetic graphs of diameter 3. *Combinatorica*
            4 (1984) 197-206.   [148]

[PS3]   (with N. Srinivasan) An extremal problem in geodetic graphs. *Discrete Math.* **49** (1984) 151-159.   [148]

            see also [NP1]

Peled, U.N.   see [HMPT1]

Peyrat, C.

[PRS1]   (with D.F. Rall and P.J. Slater) On iterated clique graphs with
            increasing diameters. *J. Graph Theory* **10** (1986) 167-171.   [115]

            see also [BBPP1]

Pierce, A.R.

[P2]   Bibliography on algorithms for shortest path, shortest spanning
            tree and related circuit routing problems (1956-1974). *Networks* **5**
            (1975) 129-149.   [272]

Plantholt, M.   see [HP4]

Plesník, J.

[P3] Critical graphs of given diameter. *Acta Fac. R.N. Univ. Comen.*, *Math.* **30** (1975) 71-93.   [104]

[P4] Two constructions of geodetic graphs. *Math. Slovaca* **27** (1977) 65-77.   [147]

[P5] On the sum of all distances in a graph or digraph. *J. Graph Theory* **8** (1984) 1-21.   [58]

[P6] A construction of geodetic graphs based on pulling subgraphs homeomorphic to complete graphs. *J. Combin. Theory* **36B** (1984) 284-297.   [149]

Plummer, M.D.

[P7] On minimal blocks. *Trans. Amer. Math. Soc.* **134** (1968) 85-94.   [104]

Pollack, H.O. see [GP1]

Pósa, L.

[P8] A theorem concerning hamiltonian lines. *Magyar Tud. Akad. Mat. Kutato Int. Kuzl.* **7** (1962) 225-226.   [78]

Price, D.G. see [CP1]

Prim, R.C.

[P9] Shortest connection networks and some generalizations. *Bell System Tech. J.* **36** (1957) 1389-1401.   [264]

Proskurowski, A.

[P10] Centers of maximal outerplanar graphs. *J. Graph Theory* **4** (1980) 75-79.   [36]

Quaif, H. see [KQ2]

Quintas, L.V.

[QS1] (with P.J. Slater) Pairs of non-isomorphic graphs having the same path degree sequence. *Match* **12** (1981) 75-86.   [185,196]

   see also [BKQ1], [BKQ2], [BHQ1], [KQ1], p. 196 and p. 203

Rall, D.F. see [PRS1]

Randic, M.

[R1] Characterizations of atoms, molecules, and classes of molecules based on paths enumerations. *MATCH* **7** (1979) 5-64. [180,196]

Rao, S.B. see [HR4]

Ray-Chaudhuri, D.

[R2]   Characterizations of line graphs. *J. Combin. Theory* **3** (1967) 201-214.   [30]

Read, R.C.   see [HKMR1]

Rédei, L.

[R3]   Ein kombinatorischer Satz. *Acta Litt. Szeged* **7** (1934) 39-43.   [222]

Reid, K.B.

[R4]   Centroids to centers in trees. *Networks* (to appear).   [54,56]

         see also [GR1]

Reingold, E.M.

[RND1]   (with J. Nievergelt and N. Deo) *Combinatorial Algorithms: Theory and Practice.* Prentice-Hall, Englewood Cliffs, NJ (1977).   [249,269]

Říha, S.

[R5]   A new proof of the theorem by Fleishner. *J. Combin. Theory* (to appear).   [86]

Ringel, G.

[R6]   Selbstkomplementäre Graphen. *Arch. Math.* **14** (1963) 354-358.   [22]

Robbins, H.E.

[R7]   A theorem on graphs with an application to a problem of traffic control. *Amer. Math. Monthly* **46** (1939) 281-283.   [207]

Roberts, F.S.

[RS1]   (with J.H. Spencer) A characterization of clique graphs. *J. Combin. Theory* **10B** (1971) 102-108.   [26]

Robinson, D.R.   see [FR1]

Robinson, R.W.   see [HR1]

van Rooij, A.

[RW1]   (with H. Wilf) The interchange graphs of a finite graph. *Acta Math. Acad. Sci. Hungar.* **16** (1965) 263-269.   [28]

Rosa, A.   see [BRZ1]

Ross, I.C.   see [HR2] and [HR3]

Rossi, P.   see [GHR1]

Rousseau, C.C.   see [FRS1]

Roy, B.

[R8]   Nombre chromatique et plus longs chemins d'un graphe. *Rev.*
            Française Informat. Recherche Oper. 1R (1967) 127-132.   [217]

Sabidussi, G.

[S1]   Graphs with given group and given graph-theoretical properties.
            Canad. J. Math. 9 (1957) 515-525.   [161]

[S2]   The centrality index of a graph. *Psychometrica* 31 (1966) 581-603.
                                                                          [44]

Sachs, H.

[S3]   Über selbstkomplementäre Graphen. *Publ. Math. Debrecen* 9
            (1962) 270-288.   [22]

            see also [ES2]

Scapellato, R.

[S4]   Geodetic graphs of diameter two and some related structures. *J.*
            Combin. Theory 44B (1986) 218-229.   [147]

Schelp, R.H.   see [EFGS1], [FGJS1], [FRS1], [FS1] and [FS2]

Schmeichel, E.F.   see [BS1], [BMSV1] and [BSV1]

Schwenk, A.J.

[S5]   Almost all trees are cospectral. *New Directions in Graph Theory*,
            Academic Press (1973) 275-307.   [132]

Seidel, J.J.

[S6]   Strongly regular graphs with $(-1, 1, 0)$-adjacency matrix having
            eigenvalue 3. *Linear Alg. Appl.* 1 (1968) 281-298.   [171]

[S7]   Strongly regular graphs. *Surveys in Combinatorics (Proc. 7th Brit.*
            Combin. Conf.), LMS Lecture Notes 38 (1979) 157-180.   [171]

Seidman, S.B.   see [ES3]

Sekanina, M.

[S8]   On an ordering of the set of vertices of a connected graph. *Publ.*
            Fac. Sci. Univ. Brno 412 (1960) 137-142.   [86]

Shannon, C.E.   see [EFS1]

Shawe-Taylor, J.   see [BBS1] and [GS1]

Sheard, M.   see [KLS1] and [KLS2]

Shier, D.   see [LS1]

Siegrist, K.T.   see [ASS1]

Simchi-Levi, D.   see [BST1]

Simões-Pereira, J.M.S.

[S9]   A note on the tree realizability of a distance matrix. *J. Combin. Theory* **6** (1969) 303-310.   [130]

[S10]   An optimality criterion for graph embeddings of metrics. *SIAM J. Disc. Math.* **1** (1988) 223-229.   [131]

[S11]   A note on optimal and suboptimal digraph realizations of quasidistance matrices. *SIAM J. Alg. Disc. Methods* **5** (1984) 117-132.
                                                             [214]

[S12]   An algorithm and its role in the study of optimal graph realizations of distance matrices. *Discrete Math.* **76** (1989) (to appear). [134]

[SZ1]   (with C. Zamfirescu) Submatrices of non-tree-realizable distance matrices. *Linear Alg. Appl.* **44** (1982) 1-17.   [131,133]

Simpson, J.E.   see [BS2]

Singleton, R.R.   see [HS1]

Slater, P.J.

[S13]   Structure of the k-centra of a tree. *Congressus Numer.* **21** (1978) 663-670.   [53]

[S14]   Medians of arbitrary graphs. *J. Graph Theory* **4** (1980) 389-392.
                                                             [45-46]

[S15]   Maximin facility location. *J. Res. Natl. Bur. Stand.* **79B** (1975) 107-115.   [34]

[S16]   Centrality of paths and vertices in a graph: cores and pits. *The Theory and Applications of Graphs*, Wiley, New York (1981) 529-542.   [47,49,56]

[S17]   Accretion Centers: a generalization of branch weight centroids. *Discrete Appl. Math.* **3** (1981) 187-192.   [54,56]

[S18]   The k-nucleus of a graph. *Networks* **11** (1981) 233-242.   [56]

[S19]   Locating central paths in a graph. *Transportation Sci.* **16** (1982) 1-18.   [48,242]

[S19a]   On locating a facility to service areas within a network. *Oper. Research* **29** (1981) 523-531.   [242]

[S20]   Counterexamples to Randić's conjecture on distance degree sequences for trees. *J. Graph Theory* 6 (1982) 89-91. [180,184,196]

see also [ASS1], [BMS1], [EJS2], [ES1], [HHS1], [MS2], [MS3], [PRS1], and [QS1]

Sloane, N.J.A.

[S21]   Hamiltonian cycles in a graph of degree 4. *J. Combin. Theory* 6 (1969) 311-312.   [83]

Smith, C.A.B.

[ST1]   (with W.T. Tutte) On unicursal paths in a network of degree 4. *Amer. Math. Monthly* 48 (1941) 233-237.   [215]

see also [BSST1]

Smith, D.H.

[S22]   Primitive and imprimitive graphs. *Quart. J. Math.* 22 (1971) 551-557.   [169,172]

[S23]   On tetravalent graphs. *J. London Math. Soc.* 6 (1973) 659-662.
[167]

see also [BS4]

Smyth, W.F.   see [CS1]

Sotteau, D.   see [BGHS1]

Spencer, J.H.   see [RS1]

Snyder, D.E.   see [EJS1]

Srinivasan, N.   see [AOS1], [AS1], [PS1], [PS2], and [PS3]

Steadman, P.   see [MS1]

Stemple, J.G.

[S24]   Geodetic graphs of diameter two. *J. Combin. Theory* 17B (1974) 266-280.   [147-148]

[S25]   Geodetic graphs homeomorphic to a complete graph. *Ann. NY Acad. Sci.* 319 (1979) 101-117.   [147]

[SW1]   (with M.E. Watkins) On planar geodetic graphs. *J. Combin. Theory* 4 (1968) 101-117.   [146]

Stewart, M.J.   see see [CS2]

Stone, A.H.   see [BSST1]

Suffel, C.   see [BS5] and [BS6]

References   **323**

Superville, L.   see [BS7]

Swamy, M.N.S.

[ST2]   (with K. Thulasiraman) *Graphs, Networks, and Algorithms.* Wiley,
               New York (1981).   [236,244]

Syslo, M.M.

[SDK1]   (with N. Deo and J.S. Kowalik) *Discrete Optimization Algorithms.*
               Prentice-Hall, Englewood Cliffs (1983). [247,258,263,268-269,279]

               see also [BS3]

Szekeres, G.

[S26]   Distribution of labeled trees by diameter. *Lecture Notes in Math.*
               1036 (1983) 392-397.   [57]

Tait, P.G.

[T1]   Remarks on the coloring of maps. *Proc. Royal Soc. Edinburgh* **10**
               (1980) 729.   [94]

Tamir, A.   see [BST1]

Tarjan, R.

[T2]   Depth-first search and linear graph algorithms. *SIAM J. Comput.*
               1 (1972) 146-160.   [235]

Taylor, D.E.

[T3]   Regular 2-graphs. *Proc. London Math. Soc.* **35** (1977) 257-274.
                                                                             [170]

[TL1]   (with R. Levingston) Distance-regular graphs. *Lecture Notes in
               Math.* **686**, Springer, Berlin (1978) 313-323.   [170,172]

Thomassen, C.

[T4]   Hypohamiltonian and hypotraceable graphs. *Discrete Math.* **9**
               (1974) 91-96.   [94]

[T5]   An Ore-type condition implying a digraph to be pancyclic. *Discrete
               Math.* **19** (1977) 85-92.   [220]

[T6]   Counterexamples to Faudree and Schelp's conjecture on hamil-
               tonian-connected graphs. *J. Graph Theory* **2** (1978) 341-347. [195]

[T7]   Planar cubic hypohamiltonian and hypotraceable graphs. *J. Com-
               bin. Theory* **30B** (1981) 36-44.   [94]

[T8]   A theorem on paths in planar graphs. *J. Graph Theory* **7** (1983)
               169-176.   [94]

[T9]   Reflections on graph theory. *J. Graph Theory* **10** (1986) 309-324.
                           [89]

     see also [BT4] and p. 115

Thulasiraman, K. see [ST2]

Tian, S. see [COTZ1] and [OT1]

Tindell, R. see [BT1], [BT2], [BT3], and [HT2]

Tomescu, I. see [HMPT1], [MT1], and [MT2]

Trauth, Jr., C.A. see [HT1]

Truszczyński, M.

[T10]   Centers and centroids of unicyclic graphs. *Math. Slovaca* **35** (1985) 223-228.   [43]

Turner, J.

[T11]   Point-symmetric graphs with a prime number of points. *J. Combin. Theory* **3** (1967) 136-145.   [75,166]

Tutte, W.T.

[T12]   On hamiltonian circuits. *J. London Math. Soc.* **21** (1946) 98-101.
                          [85,94]

[T13]   The factorization of linear graphs. *J. London Math. Soc.* **22** (1947) 107-111.   [72]

[T13a]   The dissection of equilateral triangles into equilateral triangles. *Proc. Cambridge Philos. Soc.* **44** (1948) 463-482.   [215]

[T14]   A theorem on planar graphs. *Trans. Amer. Math. Soc.* **82** (1956) 99-116.   [94]

[T15]   A theory of 3-connected graphs. *Indag. Math.* **23** (1961) 441-455.
                           [62]

[T16]   *The Connectivity of Graphs.* Toronto Univ. Press, Toronto (1967).
                         [164,165]

     see also [BSST1], [EHT1], and [ST1]

Ulam, S.M.

[U1]   *A Collection of Mathematical Problems.* Wiley, New York (1960).
                           [11]

Van Lint, J.H. see [CV1]

Vazirani, V.V. see [MV1]

Veldman, H.J.   see [BMSV1] and [BSV1]

Vizing, V.G.

[V1]   The number of edges in a graph of a given radius. *Soviet Math. Dokl.* **8** (1967) 535-536.   [56,95]

Wall, C.E.   see [CW1]

Walther, H.   see p. 112

Wang, A.Y.   see [LW1]

Wang, J.   see [BW2] and [BW3]

Watanabe, H.   see [IOW1]

Watkins, M.E.   see [SW1]

Weisfeiler, B.

[W1]   *On construction and identification of graphs.* Lecture Notes in Math. **558**, Springer-Verlag, Berlin (1976).   [169]

White, A.T.

[W2]   *Graphs, Groups and Surfaces.* North-Holland, Amsterdam (1973).   [155]

Whitney, H.

[W3]   Congruent graphs and the connectivity of graphs. *Amer. J. Math.* **54** (1932) 150-168.   [60,66]

Wilcox, G.   see [HW1]

Wilf, H.   see [RW1]

Williamson, J.E.

[W4]   Panconnected graphs II. *Period. Math. Hungar.* **8** (1977) 105-116.   [92]
            see also [AW1]

Wilson, R.J.   see [BW1]

Winkler, P.

[W5]   The metric structure of graphs: theory and applications. *Surveys in Combinatorics*, Lecture Notes in Math., London Math. Soc. **123** (1987) 197-221.   [142]

Womble, E.G.   see [KW1]

Wong, P.K.

[W6]   Cages - a survey. *J. Graph Theory* **6** (1982) 1-22.   [166]

Woodall, D.R.

[W7]   Sufficient conditions for circuits in graphs. *Proc. London Math. Soc.* **24** (1972) 739-755.   [219]

Xu, C.   see [EHX1]

Xu, Y.H.   see [HX1]

Yau, S.S.   see [HY1]

Yu, Z.G.   see [LYZ1]

Zamfirescu, C.   see [SZ1]

Zamfirescu, T.

[Z1]   On the line connectivity of line graphs. *Math. Ann.* **187** (1970) 305-309.   [64]

[Z2]   Graphen, in welchen je zwei Eckpunkte von einem längsten Weg vermieden werden. *Ann. Univ. Ferrara, Sez. VII - Sc. Mat.* **21** (1975) 17-24.   [112]

[Z3]   On longest paths and circuits in graphs. *Math. Scand.* **38** (1976) 211-239.   [112]

Zelinka, B.

[Z4]   Medians and peripherians of trees. *Arch. Math. (Brno)* **4** (1968) 87-95.   [34,42]

[Z5]   On a certain distance between isomorphism classes of graphs. *Časopis Pěst. Mat.* **100** (1975) 371-373.   [145]

Zhu, Y.J.   see [LYZ1]

Znám, Š.   see [BKZ1] and [BRZ1]

Zou, H.B.   see [COTZ1]

Zykov, A.A.

[Z6]   On some properties of linear complexes. [in Russian] *Mat. Sbornik* **24** (1949) 163-188.   [23]

Index